WA 1092239 3
D0590558

Site agent's handbook

Construction under the ICE Conditions

Tom Redshaw, BSc

Thomas Telford, London

Published by Thomas Telford Ltd, Thomas Telford House,
1 Heron Quay, London E14 9XF.

First published 1990

British Library Cataloguing in Publication Data
Redshaw, T.H.
Site agent's handbook: construction under the ICE conditions
1. Great Britain. Construction. Contracts.
I. Title.
344.103

ISBN 0 7277 1540 2

Typeset in Great Britain by The Alden Press.
Printed and bound in Great Britain by Redwood Press Ltd, Wiltshire.

Preface

This book is written with the objective of promoting a greater practical understanding between all those involved in construction projects of contracts based on the ICE *Conditions of Contract*, fifth edition—which will be referred to throughout the book as the ICE *Conditions*.[1]

The author has worked for a contractor for many years and the book is written from that experience, but some effort has been made to include points of concern to the Employer and the Engineer. There is nothing more certain than the fact that if one party to the Contract has a problem, to some extent that problem, or a consequence of it, will be passed on to the other party.

The book is essentially a non-legal appraisal of working with the ICE *Conditions* which may not always coincide with that of a lawyer. The book concentrates on work situations and for this reason the more usual method of giving full consideration to each clause sequentially has not been adopted. Instead, each chapter examines some part of the construction process and how the relevant sections of any clauses affect and control that part of the process. At the same time consideration is given to commercial pressures, whose influence on events can seldom be ignored. It is hoped that the reader will find such concentration on each subject in turn an aid to maintaining interest and to obtaining a fuller understanding of the overall effect on the particular situation. Unfortunately, the abandonment of sequential comment does have the drawback of the same clauses appearing in more than one chapter, where the content of these covers more than one subject matter. For instance clause 14 involves execution of the Works: time and extension of time; valuation of the Works and claims—all of which are the subject of separate chapters.

Throughout the book much emphasis is given to the commercial aspects of civil engineering contracts, but no apologies are considered necessary for this. The form of contract used on a variety of projects may be the same, but the circumstances which, at least initially, affect the relationship between Contractor, Employer and Engineer and the actions each take may be quite different.

With regard to civil engineering work in the private sector, instances of the Employer's being a developer, who is to sell the project on, have been virtually non-existent, but with the possibility of privately funded roads, railways and so on some examples may occur in the future. In such a case the Employer's profit will depend very much on minimising the cost of the Works to him. If he is hoping to make 10% profit on the deal, an increase in cost of 10% will wipe out his profit. However, for most private civil engineering works the Employer's profit comes from using the completed project, whose cost will only be a part of the cost of production. In some cases such as the construction of a refinery or a chemical process plant the cost of the civil engineering works can be as little as a tenth of the cost of the overall installation. In such a case the Employer's main concern will be to prevent any problems on the civil work from affecting the other 90% of work on the project.

For public works the accent has been on keeping to budget, which has involved restraining any increases in contract value, but problems can also arise with budgets if the work goes faster than anticipated, and a year's budget is exceeded. Budget problems also affect private projects and the contractor must minimise the consequences of them by keeping the Employer well informed of anticipated valuations.

In recent years, the importance of time has been recognised as being of greater importance than previously indicated in public sector motorway repair contracts. In these, the Contractor is required to pay the Employer a fixed amount for each day that the Contractor has possession of a carriageway. The Contractor allows in his tender price for the number of days of possession he needs to complete the work and he is excused payment for any extension of time granted, but he automatically compensates the Employer for any delay in completion he causes. There is therefore no need for a Liquidated Damages clause, but the daily amount inserted in these contracts has been two or three times the value of Liquidated Damages that would previously have been incorporated into a contract of that size.

For the Contractor, profit is essential to his continued existence, and this profit is directly geared to the cost of the Works to him, and the amount he is paid for them. To repeat a phrase much favoured by business consultants: 'profit depends on minimising costs and maximising value'. Whatever profit may be included in the tenders, contracting profits are probably in the region of 2–3% of turnover. It is not difficult to calculate that a half per cent increase in value raises the profit 16–25%

(one half of 3% is just over 16% and one half of 2% is 25%). The same percentage decrease in cost will affect the profit by approximately the same amount.

Some civil engineers accuse contractors of concentrating more on the maximising of value than they do on the minimising of costs and there may well be some truth in this. There is no doubt that the reduction of costs involves very much an all round effort, whereas the increase of value can be undertaken by one or two persons. However, some Engineers' antipathy towards the pursuance of small items may reflect a lack of appreciation of the relatively significant effect these have on profit. In addition the pursuit of value—in which they are directly involved—is obvious to the Engineer but the Contractor's efforts to reduce cost are not.

One should not forget that consulting firms who act as Engineers on contracts are also in a business that can only flourish if profits are made. Such firms can be just as adept as any contractor at protecting their own interests, in what has become a more difficult environment for them. Competition on fees and difficulties with professional indemnity insurance premiums and excesses are additional problems with which they have had to contend.

The book aims to marry all the differing concerns of Contractor, Employer and Engineer, and to allow all aspects to be examined before they become serious problems.

Contents

1. General

The clauses to which reference is made in this chapter are: 1, 2, 3, 5, 9, 11, 12, 13, 15, 16, 21, 23(2), 27, 35, 36, 38, 44, 48, 60, 61, 63, 66, 68, Form of Tender and Appendix to Tender.

Tenders

Most civil engineering contracts in the United Kingdom are the result of competitive tendering, either from a list of approved contractors or from prequalified contractors who reply to a published invitation to tender. The selected list of tenderers for a project may be chosen by the Employer from the lists either of approved or prequalified contractors. However the final list of tenderers is obtained there should be similar rules to limit the number of tenderers selected. It is no secret that any contractor will want the lists on which he appears to be as short as possible, thus enhancing his prospects of being awarded the job. However, unless the number of approved or prequalified contractors decreases, smaller tender lists involve the compensating situation of each contractor appearing on fewer lists. The smaller the number of tenders the greater the attention that can be given to each one, while an overall reduction in tendering costs results for both Employer and Tenderer. The average success rate per tender submitted should be the reciprocal of the average number of contractors on each list, although, in any particular period, the success rate will vary between individual tenderers.

Payments from Employers, on Contracts awarded, are a contractor's only source of income to cover his costs and profit. It is therefore Employers who foot the bill for all tenders, with the exception of those made by contractors who leave contracting to concentrate on other parts of their business or who become insolvent. Despite this more than 30 tenderers have been allowed on the list—for quite small contracts—in times of shortage of work. This reflects little credit on either the Engineer who is supervising that tendering or any contractor who remains on such a list in order to submit a tender. The instance mentioned was probably near a record, but would have resulted in a very wide range of bid prices, with a good chance that the successful tenderer would be less than ecstatic with the outcome of the Contract. Contractors would recover most of the cost of the 30 tenders from future contracts awarded. This would spread that cost on to Employers who may have maintained short tender lists.

It is therefore in the Employer's interest to obtain only as many tenders as will ensure an optimum cost of tenders while maintaining real competition. As far as the Employer is concerned the danger of a small number of tenderers is the lack of adequate competition, especially if, for some reason, one or more of those on the list are in no position to submit a really keen tender.

The cost of preparing tenders depends not only on the number of tenders, but also on the complexity of the project for which the tenders are obtained. Where the project contains a high proportion of design on temporary or permanent work the cost of a properly prepared tender is high compared with the more straightforward job. This needs to be taken into account when deciding on the correct number of tenders to obtain. The usual numbers that give optimum cost as well as competition are four tenders for projects with a significant design content and six to eight for others. Large numbers on the tender list can lead to sloppy tenders, where mistakes creep in. Although the Employer may obtain a very low tender, through a mistake, the advantage in price may prove no compensation for problems that arise during construction and settlement of the Contract.

The ICE *Conditions* is one of the documents on which the tender is based. The others, which are listed in the Form of Tender are: Drawings; Specification; Bill of Quantities; Form of Tender and its Appendix. The Tenderer's offer is made on the signed Form of Tender and not by any accompanying letter. To be strictly correct therefore the Tenderer should ensure that any other condition or document that he wishes to be part of his tender is identified on the Form of Tender.

Care is necessary if the tender is added to, to avoid the tender's being disallowed through the contravention of some rule that governs qualifications. The most usual rule is that a qualified tender will not be considered unless it is an alternative to a submitted, unqualified tender. Adoption of this rule is suggested in the Institution of Civil Engineers' booklet *Guidance on the preparation, submission and consideration of tenders for civil engineering projects*. The contractor who submitted a tender that contravened the Employer's conditions for tenders but which, despite this, was still considered, could be in a position of advantage over other tenderers. Such a situation should not occur, because the tender is, in effect, a qualified tender—which the Employer and the Engineer should, by their own rules, reject. However, the temptation to bend the rules is sometimes hard for the Employer to resist, when he may obtain a significant advantage by so doing. If the rules are bent in favour of one tenderer, others may feel they may have lost out by keeping to the rules and might consider how they too could get away with it in future.

There seems to be no hard and fast rule with regard to the acceptance of a qualified tender which is submitted with a higher priced, unqualified tender. The contractor who puts in the qualified tender will have obeyed the rules, but can find himself at a disadvantage if the Employer is interested in the alternative price, and other contractors' unqualified bids are lower than his. The Employer may consider that a request for prices from the other tenderers may result in a lower bid on the alternative basis. If this happens the original contractor will be the only one not to have the opportunity to re-price, after tenders have been opened, when there is the possibility that prices have been leaked.

It is not only qualifications that must be considered when examining tenders. Some contractors keep a look out for methods of pricing the Tender Bill of Quantities which, in their opinion, will minimise the tender total but maximise payments. This may be by way of anticipation of some necessary variation, the discovery of errors in or omissions from the Bill of Quantities, or the front loading of some prices to obtain an early cash flow advantage. A rather extreme example of the latter situation is the pricing of 'set up offices' at £1 million and 'remove offices' at a credit of £950,000.

Engineers do not care for opportunity pricing but, to a large extent, prevention is in their own hands, as opportunity is at a minimum for a well prepared and documented scheme. Opportunity pricing often arises through the Engineer's attempt to have rates in the Tender to cover a number of situations that could arise during the execution of the Works. Those Contractors who guess the reason for the items, in the Bill of Quantities, could decide to price the items on an opportunity basis. The rates inserted would give the greatest income, based on their opinion of the work that would be required. As a result the Engineer, when considering tenders, is faced with solving the problem that he could not solve when he covered a number of possible situations in the Bill. He may then decide that he would have been better off with a Provisional Sum in the Bill, rather than the multiplicity of items. The best known example of this was the Department of Transport's method of billing marginal excavation, which posed more problems than it solved and was sensibly dropped.

Quite often the Tender Documents require that the tender must remain open for acceptance for a certain period but, where this is not so, contractors should include a time limit within which the tender can be accepted. If tenders are still under consideration near the limit of time allowed for acceptance, Contractors can extend the period, should they so wish. The usually accepted time limit is two months, but this is general only and could change with particular circumstances. Reasonable time limits for the acceptance of tenders are not normally regarded as qualifications to tenders. To be strictly correct, the time limit for acceptance should be inserted on the Form of Tender, not on an accompanying letter.

The Contract

As far as a contractor is concerned, the submission of a tender is the virtual equivalent of signing a contract, because, on its unqualified acceptance by the Employer, a legally binding contract is formed between the Tenderer and the Employer. In England a tenderer has the legal right to withdraw a tender, at any time before acceptance, but in Scotland the tender must remain open to acceptance for the full period of time stated therein. Nevertheless, commercial considerations discourage contractors from withdrawing English tenders. For these reasons it is illogical, to say the least, for a company to allow someone to sign tenders who is not authorised to sign contracts.

In many cases the contract between the Contractor and Employer rests on Tender and acceptance only, with no formal Contract Agreement being either signed or sealed. Both the Department of Transport (DTp) and the Department of the Environment (DoE) rely purely on acceptance only, although the DoE does not use the ICE *Conditions*. The main advantage to the Employer of a formal Agreement is that a longer Statute of Limitations period applies, if the Agreement is sealed, rather than signed (12 years instead of 6 years). Whether the contract rests purely on acceptance, or acceptance is followed by an agreement, it is essential to be able to identify all the documents that form the contract. Ideally, any documents further to those included in the Tender should be listed in the Acceptance letter. If the contractor considers that there are any omissions, he should take immediate action to rectify the situation.

All the separate contract documents and any amendments to them should be gathered together, initialled and kept as a set, by both Employer and Contractor. The collection will perhaps be done by the Engineer, on behalf of the Employer, but only the signatures of formally authorised representatives of the Employer can validate an Agreement. The Contractor's legal, quantity surveying or estimating department may be responsible for the assembly of documents for signature by directors or the company secretary, in accordance with the articles of agreement for the particular company. The identification of a set of documents is almost certain to be necessary, if a Contract Agreement is to be signed, but can easily be overlooked where the Contract rests on Tender and Acceptance.[2]

Some care is needed to ensure that the documents initialled are those on which the tender was based. The most likely mistake to occur is in relation to the Drawings, if the design has been progressed during the period between invitation to tender and acceptance. Sometimes the negatives are amended without keeping enough original clean prints to use as identified tender drawings. Provided that the Engineer amends the suffix to the drawing number each time a drawing is amended, a check of drawing numbers against those listed in the bills of quantities or specification should avoid inclusion of the wrong documents in the contract. Any mistaken inclusion of documents in the Contract could probably be rectified later, but there may then be no reason to query the documents.

Once the Contract is formed the Contractor is committed to the construction, completion and maintenance of the Works in strict accordance with the Contract, unless released from this obligation, through the occurrence of any of the events mentioned in Chapter 7. The parties to the Contract are the Employer and the Contractor, but as the ICE *Conditions* are written, the main role played by the Employer during construction is to pay certificates, unless things go wrong.

Both Contractor and Employer should ensure that the other party to the Contract is as expected and not some subsidiary or associated company. The Contractor should be the company or person who made the offer and the Employer the authority, company or person to whom that offer was addressed. The name of the Employer and Contractor should be written in clause 1 of the ICE *Conditions*.

clause 1

The Engineer

It is the Engineer who represents the Employer in most dealings with the Contractor with regard to the construction of the Works, but the Engineer has two separate roles in the Contract

- to act as the agent of the Employer for such things as: design of the Works; provision of site services in accordance with the Specification; issuing of drawings, variations or other necessary instructions; approval of part of the Works being sub-contracted; supervision of the execution of and inspection and testing of the Works
- to give fair decisions as between Employer and Contractor whenever the Contract requires such decisions to be made; examples of such decisions are: awards of extensions of time, pricing variations, valuing claims, decisions whether or not adverse conditions met during construction could have been anticipated—the Engineer is often referred to as a quasi-arbitrator when acting in this role.

There is an argument that the Engineer acts in the Employer's interest when giving any decision under the Construction Contract, even those on fixing rates and so on, except when required to settle a dispute under clause 66. However, this does not appear to be a view widely held by Engineers and Contractors. Although clause 66 may be mentioned sometimes, or even often, in relation to claims, it is seldom invoked by Contractors. Many Engineers might regard frequent notices, under clause 66, as undesirable brinkmanship.

clause 66

clause 66

The normal commercial process for reaching settlements is negotiation and compromise, but that is not really enshrined in the ICE *Conditions*. In accordance with the conditions the Engineer comes up with his decision, after seeing and or hearing the Contractor's case and, presumably, reminding himself of the Employer's case. Both the Department of Transport and the ICE are giving consideration to conditions that will give many of the Engineer's present powers to another person, called the 'Employer's Representative'.

The Engineer may be an employee of the Employer, where for instance a local authority is carrying out some work with the technical director acting as the Engineer, or he may be a consultant engineer appointed by either a public or private sector Employer. Where the Engineer is an employee of the Employer, there may be no separate agreement that sets out his duties. It is not unknown for local rules, in contravention of the ICE *Conditions*, to inhibit the performance of the Engineer's full duties under that Contract. For example the Engineer may not be able to issue any instruction that involved amounts above a set limit without prior consultation and approval of some committee, or other official, in the authority. There may even be a series of limits that require approval at different levels. Relevant amendments to the ICE *Conditions* should therefore be incorporated into the tender documents where local rules limit the powers of the Engineer. Such amendments are in the Employer's interest in so far as they remove the Contractor's complaint that the progress of the Works was affected, to his detriment, by unrevealed limits to the Engineer's powers to act in accordance with the Contract.

If, by abiding by rules which limit his authority, the Engineer delays the issue of a necessary instruction, the total cost to the Employer is likely to be greater than it would have been without the delay. Large earthworks operations, for instance, have very high hourly costs—which do not combine well with low financial limits on the Engineer's powers to issue instructions if the aim is to minimise overall costs to the Employer. Where a consultant engineer is appointed as the Engineer, there will be an agreement between the Employer and the Engineer, under which the Engineer is paid for carrying out the duties under the Contract. The Association of Consulting Engineers publish a number of *Conditions of Engagement* which cover the duties of a consulting engineer in a variety of situations, and *Agreement 2* of this series is the one appropriate to work under the ICE *Conditions*.

Sections 6 and 7 of this agreement give details of the duties the consultant undertakes for the Employer. In particular, clause 6.3(*h*), (*j*), (*k*) and (*m*) deal with the duties of the consultant, acting as the Engineer under the construction Contract. The only limitation on the Engineer imposed by *Agreement 2* is that the consulting engineer undertakes not to give any instructions to the Contractor that in his opinion are likely to increase the cost of the Works substantially, without the prior approval of the Employer. The Contractor does not receive a copy of the consulting engineer's agreement, and will be entitled to believe that the only limitations to apply are those, if any, incorporated into the Construction contract. To avoid controversy at a later date, limitations should be incorporated into the Construction Contract to allow the Contractor the opportunity to make his representations, on any matters arising, to the person who has the power to make a decision on it.

In practice, the Engineer is not normally resident on the site, so he is unable to make the daily decisions required by the Contract, such as inspection of formations and amendment of excavation depth where necessary (clause 38). If work is to continue efficiently, the Engineer must appoint site representatives to perform the daily routine and any other duties he is prepared to delegate.

clause 38

The Contractor's agent must clarify with the Engineer the

name of his site representative, usually called the Resident Engineer, but referred to in the ICE *Conditions* as the Engineer's Representative (clause 2(1)). The Engineer's Representative's functions are 'to watch and supervise the construction completion and maintenance of the Works'. He has no power to issue any instructions that involve delay or extra payments unless such power has been delegated to him by the Engineer, in writing. The Engineer's Representative can appoint others to assist him in his function of watching and supervising, but he cannot delegate any other powers which may have been delegated to him by the Engineer (clause 2(2)). Unless the Engineer is to be almost constantly on the Site, the Engineer's Representative must be given powers greater than those conferred on him by clause 2(1). If this is not done, and the Contractor sticks to the book, frequent delays are possible, which will increase cost to the Employer.

clause 2(1)

clause 2(2)

clause 2(1)

clause 2(3)

Clause 2(3) allows the Engineer to authorise the Engineer's Representative, or his assistants, to act on his behalf in respect of any of the clauses in the ICE *Conditions* with the exception of

clause 12(3)	Decision on alleged unforeseen physical conditions or artificial obstructions
clause 44	Decisions on extension of time
clause 48	Issue of Certificate of Completion
clause 60(3)	Issue of the Final Certificate
clause 61	Issue of the Maintenance Certificate
clause 63	Certification of the value of work done by the Contractor and expenses incurred by the Employer
clause 66	Decisions on a dispute which could be a possible preliminary to arbitration.

If the Engineer decides to delegate some of his power (and it is generally to the Employer's advantage that he does) the Contractor is entitled to prior notice of this authorisation. The notice is often a separate letter to the Contractor, but there seems to be no reason why a copy of the terms of delegation to the Engineer's Representative(s) should not be enclosed. Knowledge of the actual terms would remove any suspicion that these contained undisclosed limitations—as has proved to be the case on occasions in the past—when no copy of the letter to the Engineer's Representative has been supplied to the Contractor.

On large projects there will be a need for the Engineer to delegate some powers to his representative's assistants as provided for in clause 2(3). The notice to the Contractor

clause 2(3)

should name all the assistants, the powers delegated and the part of the project to which that person is entitled to use the powers. The time to obtain the delegation should be as soon after acceptance as possible, with the matter being discussed between the Engineer and Contractor at their very first meeting. The amount of delegation should take into account the need for routine decisions to be taken every day and also the need to avoid the creation of a bottleneck by channelling too many decisions through one person.

The Contractor should be aware of any ambiguities created by delegation. For instance two people may be made responsible for taking action on interdependent clauses, such as clause 12(2) and 12(3). In accordance with clause 2(3) the Engineer's powers under clause 12(3) cannot be delegated, but those under clause 12(2) can. However, if the Engineer's Representative is allowed to make a decision, under clause 12(2) (c) to give

clause 2(3)

clause 12(2)

clause 12(3)

clause 12(2) (c)

> 'written instructions as to how the physical conditions or artificial obstructions are to be dealt with'

then he must not assume the right to instruct the Contractor on the method of working. If ground conditions were such as could have been anticipated by an experienced Contractor, the methods of construction would be up to the Contractor, who would consider that an instruction under clause 12(2) (c), specifying methods, would give him entitlement to recover, under clause 13(3), any additional costs caused.

clause 13(3)

In addition to the delegation of powers by the Engineer it is quite usual for the Employer to appoint the Engineer to act on his behalf with regard to clauses in the ICE *Conditions* where the Contractor is required to deal directly with the Employer —for example inspection of insurances under clauses 21 and 23(2), and with regard to streetworks under clause 27. If the Engineer received such an appointment, he would be acting purely as an agent for the Employer, with no role as a quasi-arbitrator.

clause 21

clause 23(2)

clause 27

Contractor's supervision

While the Contractor is pursuing the question of delegation of the Engineer's powers he has to set up his own organisation for execution of the Works. Approval for the Contractor's authorised agent who is required to be constantly on the Works has to be obtained, in writing, from the Engineer. At one time this representative was known as the Site Agent. Now the name on his door is as likely to be 'Project Manager' or on larger jobs 'Project Director'—but, for all that, he is one and the same

clause 15(2)

person (clause 15(2)). For the purpose of this book the term 'Site agent' should be taken to include project manager or director.

The Site agent is required to be in full charge of the Works and to receive instructions from the Engineer or his representatives. Unless the Engineer was resident on the Site, it would be unusual for the agent to be in a position to make every decision necessary in connection with the Works. Normally there would be a Contracts Manager or Director who would take decisions on a par with those retained by the Engineer. There is no actual condition that requires the Contractor to obtain the Engineer's written approval before the appointment of his other staff, but clause 15(1) does oblige the Contractor to employ sufficient persons of adequate knowledge

clause 15(1)

> 'during the execution of the Works and as long thereafter as the Engineer may consider necessary'.

The usual form is for the Contractor to give the Engineer or Engineer's Representative a curriculum vitae of his senior staff as they are appointed and to inform the Engineer a little before they are due to leave, on an individual basis. Information on other staff could be fed through by means of a staff return, which would also list those about to arrive or leave. If the Engineer's approval to the agent is withdrawn, his employment no longer complies with clause 15(2), and clause 16 allows the Engineer to instruct the Contractor to remove any other person, who, in his opinion, is incompetent or negligent or who misconducts himself.

clause 15(2)
clause 16

Although clause 15(2) states that the Contractor's agent should receive instructions from the Engineer or his representative, this would not be practical on larger projects, where delegation of some of the agent's duties is necessary.

clause 15(2)

Site liaison

The agent and Engineer or Engineer's Representative should arrange their respective site organisations to give the greatest flexibility, and to avoid bottlenecks, and set up systems to fit these needs. The duties and relationships between staff should be kept under review to make sure that no detrimental clash of personalities is likely to occur. Action should be taken to pre-empt such a clash, if possible, because once an incident has occurred there is a tendency for each side to give support to their own staff.

Among the items to be considered are the following.

Correspondence—the basis of the system is that letters should be addressed to the person who is authorised or delegated, under the Contract, to take the required action. Copies of all letters should be sent to the agent and the Engineer's Representative (but not addressed directly to them) to allow the maintenance of complete files, in their respective offices. There should be clear instructions issued, stating who is responsible, in each office, for ensuring that copies of letters are correctly circulated for information or action.

Instructions—Agreement to avoid verbal instructions is valuable, e.g. site instructions could be written longhand in a duplicate book and signed on the spot. The Contract does not require any instruction to be written on a particular form, but some Engineers, possibly working on their Employer's instructions, do have such a requirement. The Contractor might find that payments will not be authorised until a certain form has been issued. Being aware of this could allow the Contractor to take steps to avoid future problems, without the need for a contractual argument. The Site agent, having regard to the particular requirements of the Engineer and/or the Employer should nominate appropriate persons on his staff as authorised to receive particular instructions and to arrange confirmation of any verbal instructions received.

Meetings—Regular progress meetings, rate fixing sessions and so on should be arranged. The general principle is to avoid large numbers at any meeting by having meetings at different levels. Decisions at meetings are more likely to be achieved if maximum agreement is obtained on what the relevant facts are, before the crucial consideration of those facts. Decide who is to keep minutes of meetings, and the time limit for their circulation. It may be feasible to tape record a brief summary of any decisions made at a meeting to test that everyone present has the same understanding. The agreement of minutes can be very protracted, for people tend to remember that they said what they intended to say, or even what they ought to have said.

clause 60(1)

clause 36(1)

clause 35

Agree—a format for interim valuations, (clause 60(1)); a procedure for Inspections and for recording their approval (clause 36(1)); the form, contents and intervals at which labour returns are to be submitted, (clause 35).

Agree—Whether testing is to be joint or separate; a general testing programme.

The Site agent and head office

One of the first jobs that a Site agent should do is to arrange to visit the estimators, who were responsible for the Tender, to be briefed on the considerations built into the pricing of the work. It will probably be necessary to take particular care to understand how and why any last minute changes occurred, and their effect on the Tender rates and prices. Items affecting the work such as programme, temporary work design, concrete mix design, sub-contract analyses, placing sub-contract orders, staff, operatives and plant and material requirements should be discussed to confirm company procedures and to decide the amount of continuing work to be done on these at head office and the Site. It is particularly important that a Site agent, who is new to a company, should meet the various people in the head office that he will telephone during the course of the Contract.

2. Execution of the Works

The clauses to which reference is made in this chapter are: 1, 4, 6, 7, 8, 11, 12, 13, 14(3), 14(4), 14(5), 14(6), 15(1) and (2), 16, 17, 18, 19, 20, 26, 27, 30, 31, 32, 33, 34, 35, 36, 37, 38, 39, 42, 53, 58, 63.

Contractor's obligation with regard to Completion of the Works

clause 13(1)

In accordance with clause 13(1) the general obligation of the Contractor is

> 'Save in so far as it is legally or physically impossible the Contractor shall construct complete and maintain the Works in strict accordance with the Contract to the satisfaction of the Engineer and shall comply with and adhere strictly to the Engineer's instructions and directions on any matter connected therewith (whether mentioned in the Contract or not).'

The words 'legally or physically impossible' should not be confused with 'legally or physically difficult' or 'legally or physically expensive' as one is tempted to do on occasions. If the Contractor finds that the Works are more difficult to construct, complete or maintain than he anticipated he is still under an obligation to do so, but may be entitled to some additional payment for overcoming any difficulties which the ICE *Conditions* recognise as being appropriate for extra payment. There may also be some entitlement to recovery if unforeseen difficulties arise or because the Employer has failed in any of his obligations under the Contract.

There are a number of clauses in the ICE *Conditions* that set out the Contractor's obligations under the Contract, one of which is clause 8(1)

clause 8(1)

> 'The Contractor shall subject to the provisions of the Contract construct complete and maintain the Works and provide all labour materials Constructional Plant Temporary Works transport to and from and in or about the Site and everything whether of a temporary or permanent nature required in and for such construction completion and maintenance so far as the necessity for providing the same is

specified in or reasonably to be inferred from the Contract.'

clause 1

The nouns in the above excerpt which begin with capital letters, are all defined in clause 1. In general the method of using capitals to distinguish nouns which have a definition in clause 1, is used throughout the ICE *Conditions*.

clause 8(1)

The Contractor's obligation to construct complete and maintain the Works is, as stated in clause 8(1) 'subject to the provisions of the Contract' which would include those with regard to the timing of, the payment for, and any variations that might be required to, the Works.

The Contractor's responsibility for design

clause 8(2)

Clause 8(2) of the ICE *Conditions* requires the Contractor to be responsible for the design of the Temporary Works, but none of the Permanent Works, unless otherwise provided in the Contract. If the Contractor is to be responsible for the design or specification of some parts of the Permanent Works, then there should be some express provision to this effect, in the Contract.

clause 8(1)

Clause 8(1) also caters for the possibility that the Contractor will be required to incorporate Temporary Works, designed by the Engineer, but this is uncommon, unless the Temporary and Permanent Works designs are very closely and intricately interlinked.

What constitutes an express provision that the Contractor is responsible for part of the design can be the subject of a difference of opinion between Contractor and Engineer, unless the documents are carefully worded. For example, it is not uncommon for a Bill of Quantities to contain items for the supply of piles of a specified bearing capacity. Where this occurs, is the Contractor to assume that he is responsible for the design of the pile of the stated capacity? If the specification includes paragraphs on how the Contractor's design calculations are to be submitted for checking and so on, or the Bill item contains the phrase 'designed by the Contractor' all is clarified. Without some such further details, the Contractor may feel entitled to assume that details of the piles will be supplied on request of further drawings, under clause 7(2).

clause 7(2)

The method of billing, in this example, is not in accordance with the Civil Engineering Standard Method of Measurement,[3] so, depending on the wording of the item the Contractor may or may not have to produce a pile for the Bill rate, irrespective of who has the design responsibility.

If the Contractor is not made responsible for part of the design, at the time of acceptance, any change to the design responsibility thereafter could only be by agreement between the parties to the Contract. There is no provision in the ICE *Conditions* for the later inclusion of any such responsibility, by way of a variation under clause 51.

clause 51

Although Nominated Sub-contractors are not currently used much on civil engineering contracts, problems regarding design inclusion occasionally occur. Prospective nominated sub-contractors, being specialists, are often involved in design work with the Engineer, at a critical design stage, before tenders are invited for the civil engineering works. Such design cannot be included in a subsequent sub-contract unless an express provision is contained for it, in the Specification or the Bill of Quantities of the main contract, in accordance with clauses 8 and 58. Nevertheless, the Contractor is occasionally instructed to provide specialist work, including design, when no appropriate provision has been made in the Contract (clauses 8(2) and 58(3) and chapter 8).

clause 8
clause 58
clause 8(2)
clause 58(3)

To comply with clause 58(3) one could expect the Provisional Sum or Prime Cost Item to contain a specific reference to the design, or for such a reference to appear elsewhere in the documents. In the absence of any specific inclusion of design the Contractor is not obliged to accept a nomination, which includes design (objections to nominations are covered in chapter 8).

Unforeseen physical conditions and artificial obstructions

clause 11(1)
clause 11(2)

In accordance with clause 11(1) and (2) the contractor is deemed to have inspected the site, satisfied himself as to the ground conditions and sub-soil, access to the site and so on before submission of his tender and to have covered for any problems arising therefrom in his rates and prices, unless otherwise provided in the Contract. One such other provision, contained in clause 12, concerns the encountering, by the Contractor during the course of the execution of the Works, of

clause 12

> 'physical conditions (other than weather conditions or conditions due to weather conditions) which conditions or obstructions he considers could not have been reasonably foreseen by an experienced contractor'.

In such a case the Contractor's rates and prices are not deemed to cover the additional cost—and extra time—required to deal with the conditions or obstructions, in so far as the Engineer may agree that they could not have been reasonably foreseen. The significance of this clause to contractors, involved in civil engineering projects, increases with the proportion of earthworks in the total project. The clause has been included in civil engineering contracts because of the generally high percentage cost of earthworks in these projects, compared with work done under the building conditions. Without such a clause the responsibility for unforeseen ground conditions or obstructions would generally fall on the Contractor, and he would need to make an allowance in all tenders to cover the cost of eventually being required to deal with the unforeseeable. That would constitute a sort of self insurance scheme, which is sound enough if all contractors made such an allowance, and Employers were willing to accept the principle of part payment for other projects. However, any change of this type would tend to hurt the smaller contractors more than the larger ones, who would be better able to absorb a temporary lapse in the law of averages, with regard to occurrences.

The inclusion of the clause in the contract should tend to make the Employer more amenable to commissioning a comprehensive ground survey and report, to minimise the likelihood of unanticipated conditions or obstructions that interfered with construction. It should be cheaper to deal with any particular conditions if the method of doing so is incorporated into the original planning. Additional cost is one of the most certain consequences of uncertainty and change, because accommodation of the change can involve the cost of idle labour and plant and disruption to other items which need not have been involved.

clause 12(1)

Under the ICE *Conditions* the test of any physical conditions or artificial obstructions being deemed to be unforeseen is that they would not have been foreseen by an experienced contractor. Presumably, although it is not stated in clause 12(1) 'an experienced contractor' would have had to have foreseen the problem during consideration of the information available at the time of tender, while pricing a competitive tender, in the same tender period allowed to the actual tenderers. Failing this, the conditions or obstructions would be deemed to be unforeseen. This presumption is supported by the wording of clause 11(1)

> 'The Contractor shall be deemed to have inspected and examined the Site and its surroundings and to have satisfied himself before submitting his tender as to the nature of the ground and sub-soil (so far as is practicable and having taken into account any information therewith which may have been provided by or on behalf of the Employer) . . .'.

The important words, in the present context, are '*so far as is practicable*. .'. These words must encompass the limited time available to the Contractor, and the non-practicability of a separate soil survey done by each tenderer.

The experienced contractor would also have taken into account any information on ground conditions supplied by the Employer. The problem may occasionally arise where the information is supplied on the basis that its accuracy is not guaranteed. This condition, on the ground exploration data, may have been added to give the Engineer an each-way bet —i.e. if something in the report is not taken into account by the Contractor, the Engineer can point out that he should have done so, and if he had taken note of the report, but some other condition was encountered, he can be told that he had been warned that the report was not guaranteed.

From a practical point of view an experienced contractor can only take into account what is in the report, unless some other data indicates that the report is unreliable. In carrying out the work, the Contractor's first reaction to any conditions that were not allowed for in the Tender is that they come under clause 12, and he will continue to believe this until he is satisfied that they do not. The Contractor must ensure that he complies with the provisions for notice of his claim in clause 12, and also clause 54(4) and, at the same time, strive to overcome the condition, but complying with his general duty to minimise costs. Also, clause 12(1) requires the Contractor to take the initiative in dealing with unforeseen circumstances, by informing the Engineer of

clause 12

clause 54(4)

> '. . the measures he is taking or is proposing to take and the extent of the anticipated delay or interference with the execution of the Works.'

The steps that the Contractor can take, or propose, depend on the nature of the condition or obstruction. Where only the method of working is affected, the Contractor is fully in control of any changes required to overcome the difficulty.

For instance running sand may be encountered in the excavation, requiring the use of well-point dewatering equipment, or unstable ground may involve the use of a sheet pile to support the sides, in lieu of timbering, as allowed in the Tender. The conditions may, however, be such as to require a change to a part of the Engineer's design of the Works, as for example if unstable ground is causing slips in the permanent batters, or running sand down to and below foundation level, it is likely to cause a flotation problem for the permanent Works. If unforeseen running sand leads to a flotation it will probably cause a change to the construction methods, in addition to its effect on the design. The measures, proposed by the Contractor, would be to make the affected part of the Works safe, until the new design is available, plus a request, under clause 18, for an instruction to make any exploratory excavation necessary to ascertain the extent of the problem.

clause 12

The most common alleged unforeseen circumstances claimed under clause 12 are those that affect the execution of the Works, such as the occurrence of running sand in excavations, weak clays that require readjustment of batters or strengthening to the temporary supports, higher strength rock in tunnels affecting the rate of machine excavation, occurrence of unknown obstacles which prevent sheet piles being driven to their proper depth and existing pipes or cables in the way of new work. Some of these may be accepted by the Engineer, under clause 12(3), as genuine cases, but often the principle that they could not have been foreseen is disputed. The Engineer may be convinced that the Contractor's method of working was the main cause of the problem (e.g. failure of pumps at a critical time; leaving excavations open for an unnecessarily long period; using the wrong plant).

clause 12(3)

clause 12

By no means all unforeseen ground conditions will necessitate a claim under clause 12—some may be dealt with in measurement. Unexpected rock, or a higher or lower level of rock than anticipated, in excavation, might be accommodated in the measurement. If the method of measurement used requires separate items for rock and soft excavation, based on a definition of rock, the matter might be sorted out under clause 56. The Engineer may also include provisional sums in the Bill of Quantities, in case they are needed to deal with ground conditions that might be encountered, for such items as rock bolts or other supports, for use in tunnels, compressed air for tunnelling in wet conditions, or well-point dewatering. The detail claims procedure is part of the subject matter of chapter 6 which should be read in conjunction with this chapter.

clause 56

At the time of any incident the Contractor needs to proceed with the Works. With that in mind, if he is to succeed in a claim he must be able to demonstrate

(*a*) that the planned method was viable under the conditions known at the time of Tender
(*b*) the nature of the conditions actually encountered, and the difference from those reasonably to be expected
(*c*) that the change in method was necessary because of the changed conditions
(*d*) that the unanticipated conditions persisted for all the time the changed method was used, or alternatively that it was more expensive to change back to the original method, at any particular reversion to the anticipated conditions

(*e*) that any consequential effect on other operations were genuinely due to the unanticipated conditions
(*f*) the amount of the additional direct and consequential costs caused by the unanticipated conditions.

In carrying out the operations, the Contractor must make sure that sufficient records are maintained to be in a position to prove these six points. Expert advice may be necessary, during construction, on difficult cases of unforeseen conditions.

clause 12(1)
clause 12(2)

Following notification, under clause 12(1), the Engineer has four options available, under clause 12(2). It is obviously to the Contractor's advantage if at least some of these options are taken up at the time, or better still if the conditions or obstructions are accepted, in principle, as not reasonably foreseeable.

clause 12(2) (a)

Clause 12(2) (a) allows the Engineer to require the Contractor to provide an estimate of the measures to be taken, but without signifying agreement to the principle of the unexpected nature of the conditions or obstructions stated in the notice. The Estimate is in no way binding on the Contractor, and its purpose might merely be for guidance to the Engineer, if he was considering some change to the Works to avoid the allegedly unforeseen conditions or obstructions. Nevertheless, the Contractor could at least expect that the Engineer was not unsympathetic to the notice.

clause 12(2) (b)

A decision, under clause 12(2)(b), to approve the measures proposed by the Contractor to overcome the conditions does not constitute an admission that the conditions encountered were not reasonably foreseeable. However, the Contractor would at least conclude that the Engineer was satisfied that it was necessary to adopt the measures to deal with the conditions encountered. This could give a basis for evaluation of the claim, when the unforeseen nature of the conditions had been established. It also allows the Contractor to continue with the work.

clause 12(2) (c)
clause 12(2) (b)
clause 13(1)
clause 13(3)

Instructions to the Contractor, given under clause 12(2) (c), detailing the method of dealing with the conditions or modification to the Contractor's proposals, under clause 12(2) (b), do not automatically give the Contractor the right to recover extra time and money. Again the instruction would allow the Contractor to proceed with the Works. Any entitlement to recovery would come from clause 13(1) and (3), unless the Engineer had also accepted that the conditions or obstruction were as alleged. The Contractor would need to demonstrate that any delay or cost, caused by the instructions, could not have been foreseen at the time of tender (clause 13(3)). In the case where the Contractor was unable to prove unforeseeable conditions or obstructions, the Contractor might still obtain some redress by demonstrating that his own proposals were adequate and the Engineer's instructions added unnecessary time and/or expense. The Contractor should not forget that, even where the Engineer accepts, in principle, that the condition was unforeseeable, a considerable amount of proof will be needed on valuation, and the extent of the conditions that were unforeseen will have to be demonstrated.

Unforeseen physical conditions and artificial obstructions, which affect the Engineer's design, should be considerably easier to deal with—provided that the Engineer accepts that the design has to be changed—than those that only require changed construction methods. At least, it can be shown that the Engineer did not foresee the actual situation, and this could make him amenable to accepting that any problems of construction, arising from the same conditions, could not have been reasonably foreseen by any contractor. Once a variation is made, under clause 12(2) (d), the valuation of the work on that instruction is made under clause 52 (see chapter 6). However, clause 12 would still apply to difficulties that purely concern construction. The most common situations of this kind, and the simplest, are where the line and/or level of a drain conflicts with an existing service, and either the new or existing work has to be diverted. Much more difficult are cases where ground conditions affect the stability of embankment slopes, where subsidiary arguments—e.g. the treatment of the sub-soil during excavation and filling, rate of filling or the method and degree of compaction—may cloud the issue.

clause 12(2) (d)

clause 52

The Site and access to the Site

clause 1(1) (n)

The definition of the Site in clause 1(1) (n) is

> '"Site" means the lands and other places on under in or through which the Works are to be executed and any other lands or places provided by the Employer for the purposes of the Contract.'

This definition is often replaced by one that defines the Site as the areas shown on a particular drawing, but even where this is not the case it is normal to include in the Contract Documents a drawing that outlines the Site. This latter practice can lead to the occasional discrepancy between the drawing and the clause 1 definition. For instance the drawing may show the boundary of the Site, for a new road, returning around the abutments to a new river bridge, leaving some ground—between bridge and river—as a farmer's access. However, since the bridge deck is executed over this, it is, by the definition in clause 1(1) (n), part of the Site. If the drawing proved correct, a farmer who owned the land under the bridge would be in a good position when it came to negotiating temporary use of the land to support the bridge deck during construction.

Unlike some other forms of contract, the ICE *Conditions* do not take precedence over other contract documents. Clause 5 states that the documents are mutually explanatory, so that in the particular case of the discrepancy regarding the Site at the bridge, decisions under clauses 5 and 13(3) might well go against the Contractor, unless the drawing was unclear.

clause 5

clause 5
clause 13(3)

The Contractor is not entitled to possession of the whole of the Site on the Date for Commencement of the Works. Clause 42(1) puts limitations to the amount that is handed over, based on specific particulars in the Contract, and the needs of the Contractor's programme.

clause 42(1)

> 'Save in so far as the Contract may prescribe the extent of portions of the Site of which the Contractor is to be given possession from time to time and the order in which such portions shall be made available to him and subject to any requirements in the Contract as to the order in which the Works shall be executed the Employer shall at the Date for Commencement of the Works notified under clause 41 give to the Contractor possession of so much of the Site as may be required to enable the Contractor to commence and proceed with the construction of the Works in accordance with the programme referred to in clause 14 and will from time to time as the Works proceed give to the Contractor possession of such further portions of the Site as may be required to proceed with the construction of the Works with due dispatch in accordance with the said programme.'

If possession of the Site is not received, in accordance with this passage, the Contractor may claim any consequential extension of time and additional expenditure (chapters 3 and 6).

With a green field Site, the most usual constraints on the possession of parts of the Site are problems in purchase of the land, time allowed for farmers to harvest crops or for wildlife to breed and the need to carry out any accommodation works for original owners before use can be made of the purchased land. If the accommodation works are to be carried out by the Contractor, then the constraints should be written into the Contract. If they are to be carried out by the Employer's workmen, or under separate contracts direct to the Employer, any work which is expected to intrude into the Contract Period should be noted in the Contract. This note may be a late addition to the Contract, between tender and acceptance, if the accommodation works were subject to delay, after the tender documents were prepared.

Where the Works are the alteration or renewal of an existing plant, the possibilities for constraints to the possession of parts of the Site are much more likely. This is particularly so if the existing plant is to be kept in full, or limited, operation during the construction of the Works. Such constraints could be incorporated into the Contract through the use of Sectional Completions. In some areas of archaeological importance, there may be some period, following demolition and during excavation, which is subject to disruption from inspection of arisings.

The Contract provides for the Employer's workmen, or his other contractors, to be working on the Site during the construction of the Works. The Contractor is required, by clause 31(1) to afford all reasonable facilities for them to do so. Part of these facilities is the provision of information about occupation of parts of the Site, which may be required by both the Contractor and other workmen, during the construction period, and on what stage of the Works the other work is to be carried out.

clause 31(1)

Clearance of the Site on Completion

Clause 33 requires the Contractor to clear the Site on Completion, and leave the whole of the Site clean and in a workmanlike condition to the satisfaction of the Engineer. Failure to remove any Plant, goods or materials from the Site, within the reasonable time allowed by the Engineer, gives the Employer the right to sell any which are the property of the Contractor and return to the owner, at the Contractor's expense, any which are not (clause 53(8)).

clause 33

clause 53(8)

Other land required in the execution of the Works

The Site does not necessarily constitute all the land required to carry out the Works. This is recognised in clause 42(2), which requires that

clause 42(2)

> 'The Contractor shall also provide at his own cost any additional accommodation outside the Site required by him for the purposes of the Works.'

Such other accommodation might include land for disposal of surplus material or soil which under the Contract was the property of the Contractor and for offices, workshops, precasting yards and batching plants.

It is to be hoped that the estimators will have tied up the use of any absolutely essential extra land, by option, during the tender period. If not the Contractor may suffer the consequences of a sudden, local hyperinflation in land rents. All agreements to use additional land should be drafted properly, in specific terms, with the use of specialists if necessary. Many problems in the past have occurred through the use of loose terms such as 'we will fence, drain or topsoil the area before handing the land back'. These should be avoided and replaced by exact descriptions of the work to be done, including the type of fence to be erected during the Contractor's use of the land. Generalities, to the landowner, will entail considerably more than the Contractor intends. The possibility of overrun should not be forgotten. This contingency could be covered by an option to renew. The terms for renewal should be included in the option document.

All land to be used should be fenced off from adjoining land as soon as possible, preferably before any other work is commenced. Bulldozer drivers seem to be fatally attracted to making a circular tour of any part of a field which they should keep off. Another point of concern to the landowner or tenant is that retained portions of fields cannot be used for grazing animals until fences are erected.

If at all possible materials should be stored on part of the Site, and not on other land procured by the Contractor. The reason for this is that their value can be included in interim certificates only if they are so stored, unless the materials are listed in the Form of Tender (Appendix) (clause 60(2)). The Engineer has no discretion to allow interim payments for materials stored off Site, even if the Contractor undertakes to vest property in those materials in the Employer, as he is required to do, in respect of materials on Site (clause 53(2)). The Employer's agreement is necessary, and there is a greater possibility of obtaining this before the materials are positioned than after the Contractor has committed himself (see also chapter 5, p. 104).

clause 60(2)

clause 53(2)

Access to the Site

The Conditions do not state, specifically, that the Employer will provide the Contractor with access to the Site, but this is implied by clause 42(2), which, inter alia, states 'The Contractor shall bear all expenses and charges for special or temporary wayleaves required by him in connection with access to the Site'. The converse of this requirement is that the Contractor is not responsible for ordinary or permanent wayleaves. However, access to the Site is one of the items about which the Contractor is deemed to have satisfied himself, before submitting his tender (clause 11(1)).

clause 42(2)

clause 11(1)

If, for instance, the Employer has a right of way over private roads to the site, the limitations of use may not allow the density of traffic, or the weight of individual loads, which are necessary for construction. It would be reasonable to expect such a limitation to be mentioned in the Contract, and allowance to be made at the time of tender. The Site agent would need to be aware of any such problems, in order to make arrangements in good time to avoid delay in proceeding with the work.

clause 30

Clause 30 deals with access to the site over public roads, and requires the Contractor to

> '. . . use every reasonable means to prevent any of the highways or bridges communicating with or on the routes to the Site from being subjected to extraordinary traffic within the meaning of the Highways Act 1980 or in Scotland the Road Traffic Act 1930 or any statutory modification or re-enactment thereof by any traffic of the Contractor or any of his sub-contractors . . .'

The clause goes on to require the Contractor to select routes, distribute loads and traffic and limit loads so as to ensure, as far as reasonably possible, the avoidance of unnecessary damage. In addition to this general obligation, on the part of the Contractor, to select routes and limit loads and so on, there may well be some specific requirements in the Contract limiting, or prohibiting the use of certain roads. The Site agent should not forget that such specific limitations do not replace the general obligation, unless so stated.

Apart from contractual obligations, the danger, noise, dust, disturbance and mud on roads which can arise when large quantities of materials—particularly filling materials—are brought on to the Site, constitute potential causes of friction with the local population and police. Problems are easier to solve if there is some early liaison, and this might include the local press. The treatment of any reports of incidents can make a big difference to the size of problem they cause.

The largest and heaviest loads needed on site could well be the plant, equipment and Temporary Works required for construction of the Works. The effect on roads and bridges of the transport of these to the site is dealt with in clause 30(2) and the consequences for the Contractor are different from those where materials are involved. The Contractor is required to be responsible for the cost of any necessary strengthening of roads and bridges, which can only be done through the relevant highway authority and any other parties over whose property bridges are built. The Contractor also has to idemnify the Employer against any claims for any damage caused by the

clause 30(2)

passage of plant, equipment and Temporary Works on the way to Site, which makes the Contractor responsible for payment of any claims received.

clause 30(3)

As a contrast, where materials are concerned, the Contractor is entitled, under clause 30(3), to be indemnified against any damage to roads or bridges as a consequence to their haulage to Site, provided that the haulier is not made responsible under an Act of Parliament. The Contractor must fulfil his obligations to do everything reasonably possible to avoid or reduce such damage (clauses 30(1) and 30(3)). It would appear from the three parts of clause 30, read together, that any strengthening of roads or bridges for the passage of materials or fabricated items would not be included in the reasonable precautions to avoid damage that a Contractor is required to take.

clause 30(1) clause 30(3)

Contract drawings

The Contractor is entitled to two copies of the drawings and other Contract documents free of charge, but clause 6 also allows the Contractor to obtain or make, at his own expense, any further copies he may require. The copyright of the drawings and specification remains with the Engineer.

The drawings listed in the tender specification or the Bill of Quantities are important to the Contractor, as they are the ones on which the Tender was based. One set of these should be kept in the office and not allowed to be taken out of that office (as recommended in chapter 1). The Contract does not provide for the Contractor's having all the necessary drawings at the Date for Commencement, but, by clause 7(1), the Engineer

clause 7(1)

> '. . . shall supply to the Contractor from time to time during the progress of the Works such modified or further drawings and instructions as shall in the Engineer's opinion be necessary for the purpose of the proper and adequate construction completion and maintenance of the Works . . .'

The use of the mandatory 'shall' in this clause may seem to imply that this should be done without any further action on the part of the Contractor, but clause 7(2) requires that

clause 7(2)

> 'The Contractor shall give adequate notice in writing to the Engineer of any further drawings and specification that the Contractor may require for the execution of the Works or otherwise under the Contract.'

What constitutes 'adequate notice in writing' may be contentious in the case of any alleged delay, and the Contractor should take steps to avoid arguments concerning this, which, in the absence of certainty, might otherwise become necessary. Dates on the programme issued to the Engineer under clause 14, showing the latest date for receipt of all information for certain parts of the Works, may be 'notice in writing' but a list of those dates contained in a letter to the Engineer is, beyond doubt. It would seem from the end of clause 7(3) that a drawing showing variations is deemed to be issued pursuant to clause 51, which requires variations to be in writing, and, perhaps, establishes the issue of a drawing as notice in writing. However strongly one may feel that the programme is notice in writing, it is better, particularly when so little effort is involved, to make certain by writing a letter. A fairly common, and effective, practice is to submit an up-dated schedule of information for consideration at each Site meeting.

clause 14

clause 7(3)

clause 51

Whether the list of latest dates (for the receipt of drawings or specifications) constitutes adequate notice in the context of clause 7(2) is also arguable. Where the dates are given with a programme, issued in accordance with the Contract, the notice is as long as possible and, as far as varied drawings are concerned, is the best that the Contractor can be expected to do. However, it has been argued that clause 7(2) requires more specific reference to those individual drawings or specifications that contain insufficient information to construct that part of the Works to which they refer—but that seems to go too far. Although clause 7(1), in setting out the Engineer's obligation, refers to 'such modified or further drawings and instructions' the Contractor is required, by clause 7(2), to give notice only of 'any further drawing or specification' which (bearing in mind the difference in wording) would not include an amended reissue of an existing drawing or specification—i.e. drawings showing variations. The Contractor should bear in mind that drawings issued under clause 7(1) are variations only if they comply with clause 51(1) and that Bill rates can apply to variations. A commonsense approach is necessary, from both Engineer and Contractor, to ensure the availability of drawings when required for the Works to progress. This is particularly so when much detail design has been left to be completed during the construction period.

clause 7(2)

clause 7(2)

clause 7(1)

clause 7(2)

clause 7(1)

clause 51(1)

The Contractor should notify the Engineer of any incomplete drawings, errors and discrepancies on or between drawings and specification and so on as soon as discovered. However, the ICE *Conditions* put no obligation on the Contractor to check the Engineer's drawings. A difference between an overall dimension of a structure and the sum of its components is an example of the type of error which occasionally occurs on drawings. At the time of setting-out the structure the Contractor would have no need to take the individual component dimensions into account. If these were correct and the overall dimensions were wrong, the setting-out would be wrong but based on incorrect data supplied by the Engineer. The Contractor cannot be held responsible if errors or discrepancies are not discovered in time to avoid delay and/or abortive work (clauses 5, 13(3) and 17).

clause 5
clause 13(3)
clause 17

Another example of a discrepancy appearing in the documents is the description of an item which gives a load requirement for a component, but also quotes a particular manufacturer's product that does not satisfy that requirement. This would need to be decided by the Engineer in accordance with clause 5 and an instruction would be given, under clause 13. The Engineer might decide that an experienced Contractor should have noted the discrepancy at the time of Tender and asked the Engineer to correct it, before submitting his Tender. The details available at the Time of Tender might be the deciding factor, but the Contractor could not reasonably be at fault if in his Tender he had allowed for the component's compliance with the stated load requirements.

clause 5
clause 13

It is highly important to maintain rigorously an up-dated drawing register, listing all drawings by number and amendment reference, the date received and the distribution of all copies. All copies of the previous amendment should be collected and clearly marked 'superseded' when the up-dated copies are distributed. This avoids instances of working and superseded drawings occurring in the same pile on an engineer's desk. The persons who perform the measurement will require copies of all superseded drawings and should inform the Engineer, as soon as possible, of any items of work prepared or done which becomes abortive through the issue of revised drawings. This is particularly important where the Engineer might be unaware of such items, e.g. the ordering, preparation and manufacture of materials or Temporary Works.

The Contractor is entitled to make a claim for extension of time, under clause 44, and any additional cost, under clause 52(4) (b), caused by a late receipt of a drawing or specification, which he has requested (clause 7(3)). Disputes sometimes arise as to the amount of information that the Engineer should supply to the Contractor and the manner of its presentation. One practice abhorred by Contractors is when the Engineer issues amendment notes, instead of amended drawings, and leaves the Contractor to transfer these to the relevant drawing. One or two instances of this can be tolerated, but drawings

clause 44
clause 52(4) (b)
clause 7(3)

with large numbers of changes on them become difficult to work with.

An occasional disagreement on the amount of information to be made available concerns bending schedules. These are normally prepared and issued by the Engineer, because they are needed by him to calculate the quantities of reinforcement to be billed and to be included in the final measurement. The Contractor can usually get by without bending schedules at the Tender stage, but definitely requires them when ordering cut and bent reinforcement. If the tender Bill of Quantities contains detailed reinforcement figures, tenderers assume that bending schedules will be made available when required, but this is not always the case.

In instances where bending schedules are not made available, the Engineer can argue that they are unnecessary for construction, provided the reinforcement drawings are sufficiently detailed for the Contractor to draw up schedules. If a dispute occurs it is usually because ordering of reinforcement and concreting is delayed by the need to produce schedules at short notice. However, the Contractor should not prolong that delay by ineffective argument. The best course open to him is to state his case, produce the schedules, and if convinced that right is on his side, make a claim under clause 7(3). If bending schedules will be required immediately after acceptance, and there is any doubt about availability, the Contractor should ask the Engineer for clarification, during the Tender Period. Clause 7(3) applies only to late receipt of drawings requested by the Contractor, and not to late variation drawings that cause delay and extra expense.

clause 7(3)

If the Contractor has been warned not to work to existing drawings, because some variation is being prepared, and delay occurs after the Contractor has stated the required receipt date for the varied drawings, clause 7(3) would apply. There is the alternative of regarding the instruction not to work to existing drawings as a suspension order under clause 40 of the *Conditions*, but to some extent this clause can give rise to arguments under clause 40(1) (c), which would not be so obviously applicable to clause 7(3).

clause 7(3)
clause 40
clause 40(1) (c)
clause 7(3)

Setting-out the Works

The Contractor is responsible for the correct setting-out of the Works, from the data supplied by the Engineer (clause 17). It is for the Engineer's Representative to decide how much checking of the setting-out should be done, if any, and whether this is to be achieved by joint working or by a separate exercise. However, clause 17 makes it quite clear that no responsibility

clause 17

for any errors that may have been missed in the check falls on the Engineer's Representative. The Contractor is entitled to sufficient data to enable him to set out the Works, and provided this is supplied to him, in writing, the Engineer is responsible for the correctness of such data. It is prudent for the Contractor's setting-out Engineer to have a duplicate book with him to make sure that he obtains all data in writing, especially any found to be missing at the last minute before setting out. Although the Engineer's check on setting-out removes no responsibility from the Contractor, it should always be treated as an insurance, by ensuring, as far as possible, that all calculations are done independently so as to reduce the possibility that both persons will make the same mistake. The Site agent should ensure that the Site datum points, for line and level, are fixed and separately checked by different members of his staff.

The data to be supplied by the Engineer can be co-ordinates or base lines either on the Site or remote from it. The more remote the data the greater the chance that some sort of trial and error adjustment to the data supplied may be required to fit the Works on to the Site. The Contractor should make sure that, in helping in this, he does not increase his vulnerability should errors be found later.

clause 17

Clause 17 also requires the Contractor to protect all bench marks, setting out pegs and so on. In doing so it should be remembered that there is a Murphy's Law, which states that even if it is impossible for construction plant to hit setting-out pegs, at least half of them will be hit.

Materials and workmanship

The Contractor has an obligation to supply materials and execute work in compliance with the specific provisions of the Contract and also, in accordance with clause 13(2), to the Engineer's approval. This latter requirement should not normally add to the cost of materials or workmanship. In most cases, any significant requirement over and above the drawings and specification would constitute a variation. The main problem areas, with regard to 'Engineer's Approval' are likely to be in connection with the appearance of such items as brickwork (particularly where multi-coloured bricks are used) and bushed hammered concrete.

clause 13(2)

The Contractor is as responsible to the Employer and/or Engineer for the materials supplied and workmanship executed by Sub-contractors, as he is for those within his direct control (clause 4). If a Sub-contractor does not comply with the Sub-contract (which should be same as not complying with the Contract) then he will have to correct this at his own expense,

clause 4

provided he remains solvent. Nevertheless, even where the Contractor is protected from financial loss by the Sub-contractor, his reputation with the Employer and/or Engineer will depend on his avoiding too many problems with workmanship and materials, and he, himself, should ensure that Sub-contractors maintain the proper standards.

clause 36(1)

The Contractor is required by clause 36(1) to

> '. . . provide such assistance instruments machines labour and materials as are normally required for examining measuring and testing any work and the quality weight or quantity of any materials used and shall supply samples of materials before incorporation in the Works for testing as may be selected and required by the Engineer.'

It is normal for the testing equipment to be listed in the Specification, but even if this is not done the Contractor is not exempted from supplying the normal equipment such as cube moulds and density testing equipment. However, if the Bill of Quantities has been prepared in accordance with the Civil Engineering Standard Method of Measurement,[3] the General Items section should contain items for this supply (item 2 of class A of the CESMM).

clause 36(2)

The samples used to demonstrate that materials are in accordance with the Contract are supplied at the Contractor's own cost, if the supply thereof is clearly intended by, or provided for in the Contract (clause 36(2)). Otherwise the Employer pays the cost of samples. The Contractor should make sure that any samples of materials that are provided are truly representative of the materials that will be supplied, in bulk, during the course of construction. If there are to be any problems with materials' compliance with the Contract, it is likely to be less expensive if these are brought out into the open —before work begins in earnest. Testing of materials is often concentrated mainly on the higher grade ones to be used in concreting or asphalt, but the bulk filling items should not be ignored, especially those used in foundations. Such problems as the swelling of steel slag as it reacts with moisture, or the incorporation of industrial waste which has been mixed in with colliery waste used as hardcore, can turn out to be expensive. These types of material can be variable, and frequent inspection of the source is advisable. Any materials brought on to the Site that are not in accordance with the Contract are the Contractor's responsibility, even though this has occurred in the presence of a representative of the Engineer. The possible exception is where the source of the materials has been specified by the Employer, but even so the detailed requirements of the Contract will apply to the supply. If the materials do not comply, the Employer would need to give an alternative source.

Samples of workmanship, incorporating the proposed materials, are often required. Even where this is not the case, the Contractor may find it of advantage to do his own trial run, to test the workability of a concrete mix or the possible speed of sawing joints in concrete paving. Any trial can be of little use unless it simulates the actual conditions to be encountered in the bulk production. A specimen brick panel painstakingly built from selected bricks, on a beautiful day, on a hardstanding outside the offices, taking four or five times the number of hours allowed in the estimate, may prove to be impossible to reproduce in the general working situation. Nevertheless it could, and often will, be held out as the approved standard of workmanship to be achieved. It is better to meet the problem of standard of workmanship at the sample stage rather than when full-scale operations are under way. Some increase in output will occur when operatives have had a little experience, so one would expect a little longer time to be expended on a sample, but otherwise the work should be carried out in average conditions, with average operatives and with no extra activities.

Tests to demonstrate whether or not the materials and/or workmanship comply with the Contract are at the Contractor's cost, if it is clearly intended by or provided for in the Contract (clause 36(3)). The Civil Engineering Standard Method of Measurement (CESMM)[3] requires items to be provided for many of such tests. However, any test demonstrating non-compliance of materials or workmanship is at the Contractor's cost (clause 36(3)). Tests to demonstrate the adequacy of the Engineer's design are only at the Contractor's cost if they are clearly intended by or provided for in the Contract and are particularised in the Specification or Bill of Quantities, sufficiently for them to have been priced at the time of tender (clause 36(3)). This type of test would normally be a bill item either in the general items or in the case of a load test on piles in the piling section (items 2.5 and 2.6 of class A and item 8 of class Q of the CESMM).

clause 36(3)

clause 36(3)

Under the ICE *Conditions* the Engineer is entitled to inspect and measure all work, before it is covered up. The Contractor is not allowed to proceed on to the following operation without either the Engineer's approval of the work done or his notification that inspection is unnecessary (clause 38(1)). The time work will be available for examination must be reported by the Contractor to the Engineer, who must take action without unreasonable delay.

clause 38(1)

Clause 38(1) stipulates that the Engineer is responsible for issuing approvals, but this is normally delegated to the Engineer's Representatives or an assistant (clause 2(1) and 2(2)). The Engineer's failure to delegate his powers under clause 38(1) would impose severe delays in executing the Works. Clause 38(1) does not require the Engineer's approval, or advice that inspection is unnecessary, to be in writing, so the degree of formality given to the compliance with that clause can vary with the personalities involved. However, the application of clause 38(2) will be formal, because of the possibility of contention, so a record of inspections and so on should be maintained. The Contractor's agent and the Engineer's Representative should agree, and confirm to each other, the names of those in each team responsible for compliance with clause 38, for each part of the Works, and the degree of formality to be observed. If everything is to be done verbally, the Contractor should record the date, time, item of work and who approved or advised that no inspection was necessary, for that item. If considered necessary by the Engineer, a form should be devised that gives the same information, which should be signed by both the Contractor's and the Engineer's Representative, or authorised assistant.

clause 2(1)
clause 2(2)
clause 38(1)
clause 38(2)
clause 38

The Site agent should not allow work to be inspected which is incomplete, insecure or dirty as this may well lead to the delay associated with a further inspection. Repeated obvious faults in work inspected will lead to a loss of confidence in the Contractor, and tend to lead to more stringent standards being set. The attitude towards inspections, on the part of the Contractor, that gives the least disruption and bad feeling, is one which demonstrates that they are a pure formality. The Contractor who makes up a claim for payment consequent on delays in, or unduly high standards of, inspections is being unrealistic in expecting any payment, even if there is some substance in the claim. The task of convincing the Engineer, while work proceeds, that these delays or standards are unreasonable must be considerably less than that of persuading him to certify what would be a waste of the Employer's money. Therefore, the effort should be concentrated on avoiding the need for a claim.

If work has been covered up without the Engineer's consent, the Engineer can have it opened up for inspection, and reinstated, at the Contractor's cost (clause 38(2)). Although, for instance, one would not expect an order to remove concrete from a pour, because no opportunity to inspect the reinforcement or cleanliness had been given, it is a possibility. Any opening up required, after compliance with clause 38, is only

clause 38(2)
clause 38

at the Employer's expense if the work is shown to be in accordance with the Contract. In the end the acid test of any work is its compliance or non-compliance with the Contract, not whether or not it has been approved (clause 39(1) (c) and 39(3)). Clause 39(1) and (2) empowers the Engineer to order in writing the removal of any materials or workmanship not in accordance with the Contract and their replacement with complying items. If the Contractor is in default on the order, the Employer can have the work done by others, at the Contractor's expense. In serious cases default could lead to determination (clause 63 and chapter 7).

clause 39(1)

clause 39(3)

clause 39(2)

clause 63

The Engineer's order to remove defective workmanship or materials will be based on his opinion, which may be proved to be wrong. If the Contractor disagrees with the Engineer's order he can either decide to stand and fight at the time or he can register his disagreement, obtain the details needed to prove his point and, if the Engineer remains unconvinced, remove the offending item. He can activate the disputes procedure—then or later on. If the order was made by the Engineer's Representative, the Contractor can request the Engineer to review the decision, as he is entitled to do under clause 2(4). In such a case the Contractor should ask to present his case in the presence of the Engineer's Representative, to ensure any counter arguments are fully answered, direct to the Engineer.

clause 2(4)

A disagreement between the Engineer and Contractor on whether or not an item complies with the Contract may depend on many things, but it is important to establish, and understand, the true reason for condemnation before any action is taken.

If there is a difference between Contractor's and Engineer's tests, the obvious answer is to take some additional tests on an agreed procedure. However, the Engineer may be reluctant to do so unless the Contractor can point out something irregular in the original test procedure. This is particularly likely if the amount involved is considered fairly insignificant. The Contractor and Engineer should bear in mind the extra cost that is incurred because of the disagreement, and should choose the course which minimises this. Any delays can increase costs greatly when extensions and Liquidated Damages enter into consideration. If the Engineer cannot be convinced before work is removed, and reinstated, it will take considerably more evidence for him to accept that removal was a waste of the Employer's money, for which he was responsible.

Checks on materials usage

For many operations the largest element of cost is that for materials, which is a very good reason for keeping a check on the amount and cost of the materials purchased against those certified. The amount certified for materials will include temporary materials, for which there may be no specific item in the Bills of Quantities, but which have been allowed for in the estimate, and quantities for wastage, built into the Bill rates and prices. The reconciliation of materials used against those in interim accounts is popularly referred to as a 'wastage report', which can be misleading. The real intention behind it is to show that measures being taken on Site to avoid wastage are successful, rather than to determine the level of wastage. The Site measures, which apply every day, are thus more important than the report, which is usually compiled once a month, after the work done for the month has been valued. If the daily site procedures are not followed, wastage is almost bound to occur.

When wastage is shown up on a report, it may be that there is true wastage or that the measurement may be too small, or a combination of both. If it is known that measures to prevent wastage are being rigorously applied then to some extent the report is a check on the validity of the valuation. The basis of any action taken to correct the situation should include an immediate examination of the preventative procedures, as well as of the measurements. Of the two choices of action available the former should have the higher priority. Any real wastage that has occurred will be repeated, and lost to the Contractor, but any incorrect measurement items can probably be put right.

Wastage can occur from a number of causes including

(*a*) poor storage of materials on the Site, resulting in losses due to: breakages; being run over by plant; covered over; discoloration

(*b*) poor tolerances on workmanship: e.g. formation too low; no shuttering used to contain concrete; trenches too wide for backfill materials; previous layer of hard standing material too low

(*c*) over-ordering of materials, such as concrete that cannot be stored, because of inaccurate premeasurement; the results can be seen in heaps of concrete close to pours

(*d*) paying for more materials than are delivered

(*e*) compacted densities being greater than anticipated

(*f*) theft and vandalism, particularly in urban areas

(*g*) weighing gear, such as that in a concrete batching

plant, going out of order during use

(*h*) work being broken out and replaced, because the materials or workmanship used were not in accordance with the Contract.

A fairly constant source of error, giving a false impression of the amount of wastage, are delivery tickets remaining in the ganger's pocket for some time after he has signed for the materials.

Some idea of the measures that can be taken to minimise wastage are given in the following example of mixing and laying concrete in a hard standing.

At the batching and mixing plant

- Arrange to have a single entrance/exit to ensure all material is unloaded in the stockpile areas or cement silos.
- If no weighbridge is installed on the site arrange to take sample vehicles for check weighing, both full and empty.
- Check vehicles delivering sand for excess water, and check weigh any found.
- Carry out spot checks that materials delivered comply with the specification.
- Check the proportions of the approved mix, against that assumed in the Tender, so that an appropriate allowance can be made in the reconciliation report, if necessary.
- At the end of each day dip the cement silo, and calculate the amount of cement in it. Compare this with the theoretical contents from the previous dipping, the amount delivered during the day, and the number of batches used times the weight of cement per batch (this checks that the weigh gear remains in order).
- Keep a schedule of the number of, and order in which, concrete delivery lorries leave the plant to compare with the number and order in which they arrive at the laying plant.
- Calculate the amount of concrete dispatched against the theoretical amount of concrete laid.

At the hardstanding

- Keep a check on the levels of the sub-base on which the concrete is to be laid, and correct as necessary, a day or so in advance of concreting.
- Precalculate the amount required each day and inform the batching plant of the quantity, and of any up-date of the calculated amount. This may avoid the wastage of the odd load, but will also alert someone to possibility of losses, if the quantities are changed.
- Keep a record of the number and order in which delivery

lorries arrive at the laying point, particularly if lorries have to leave the Site, during transportation of the concrete.

In any exercise to keep records the person making the measurements and observations should be completely informed of the reasons for each part of the record, so that he does not do the wrong thing. For instance dipping a cement silo may involve a hard climb up a ladder, and the dipping may be inaccurate. Later, however, instead of dipping the silo, the information on number of batches produced and amount of cement delivered may be used to calculate the height of cement in the silo. Should this occur when the weighing gear goes out of order, there may be the embarrassment of running out of cement, when, theoretically, the silo should contain 100 t.

It is to be hoped that any such change in emphasis in the records, would be discovered before the supply of cement ceases, in the middle of a pour, when the last check indicated that there should be a stock of 100 t in the silo.

Sometimes the checks are necessary to ensure that sufficient material is used, rather than to prevent excessive usage. Where, for instance, concrete is difficult to place, such as round a pipe in a heading, some may be omitted. If the amount required is calculated, and sent down the shaft, it should be placed, and the possibility of later subsidence reduced. Other measurements may also avoid problems by giving some warning of the unusual happening. The Site agent should encourage his staff to be aware of what they should see, to compare with what is actually seen. One example could be of calculating and keeping a mental picture of the amount of material arising from a bored pile, in sand. More arisings than expected may be a warning of the undermining of an adjacent building, before any real damage is done. This could result from a combination of wet sand and the pile casings not being driven low enough.

Labour and plant

clause 8(1)

Under clause 8(1), the Contractor is required to provide all labour and constructional plant necessary to construct, complete and maintain the Works. The Contractor is required to supply a return of labour and plant to the Engineer, in the form and at the intervals he may require. The amount of detail required on this return varies, depending on the Engineer or his Representative, who usually has the delegated authority to deal with this (clause 35).

clause 35

A fairly usual requirement is for a weekly return, showing daily numbers of workmen, including Sub-contractors', of each trade or category. The plant details are the numbers of each type, often with the number of hours worked for each. On the larger sites the return would probably be detailed separately for each location or structure.

Labour

Much of the work done on Site may be carried out by sub-contractors, but the Contractor is as responsible, to the Employer, for their actions as he is for those of his own workmen (chapter 8). Provision of labour on a piece work basis is not considered to be sub-contracting (clause 4). Those employed in the execution of the Works are required to be careful, skilled and experienced in their trades and callings, and the Engineer has the right to require the Contractor to remove from the Works anyone who is incompetent, or misconducts himself (clause 16).

clause 4

clause 16

The Contractor, by clause 35, is required, where appropriate, to pay rates of wages and observe hours and conditions no less favourable than those established in the Working Rule Agreement of the Civil Engineering Construction Conciliation Board for Great Britain.

clause 35

Incentive bonuses

Incentive bonus schemes have an important role in many civil engineering projects, and are particularly useful where much plant is employed. There are many variations, but the basic idea is to set the operatives a target to achieve, and to pay some part of the Contractor's cost saving if the target is exceeded. The target can be expressed as a number of units of output with the bonus being a fixed amount per unit above the target, or the bonus per unit can increase when secondary targets are reached to give an increasing incentive to achieve the higher rates of progress.

One alternative way of representing a target is to set the target as man-hours per unit, giving some proportion of the man-hours cost saved as bonus. For example if 35 units of work, each worth 2 man-hours, were executed by an operative in a week of 47 h, 23 h would have been saved and available for bonus. If the whole of the saving were to be paid to the operative he would be paid 70 h, and if 50% of the saving was paid in bonus, 58·5 h would be paid. From this it can be seen that, where only a part of the saving is to go to the operatives, the target number of hours allocated to the task must be larger, to allow the same bonus to be earned, in both cases.

Incentives are more effective if each person's bonus depends solely on his individual effort, but this is rarely possible, and

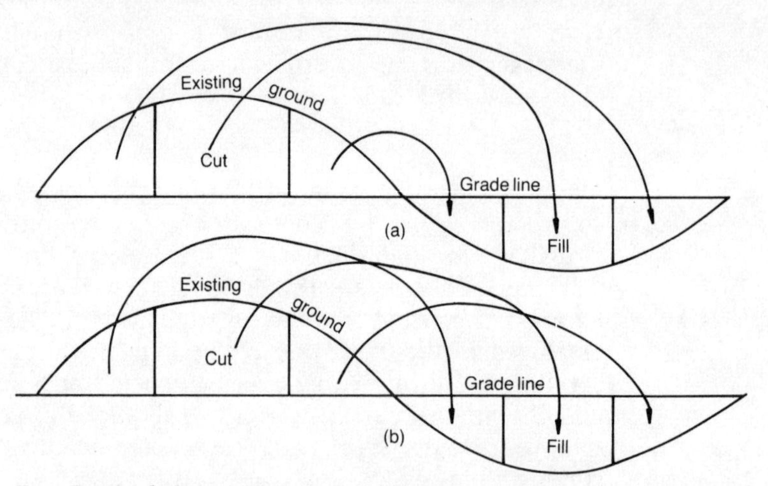

Fig. 1. Haul distances for excavation and filling: (a) varying; (b) constant

bonuses relate to teams.

Another method of bonusing for a team is to set a target of man-hours for a specified operation, which may last several days but should be less than a week. Each member of the team would have a share of the total bonus earned, on the basis of their individual importance to the overall performance. Share values are not necessarily proportional to the hourly rates of pay of the individuals, but are fixed to try to give the greatest incentive to the person who is able to set the pace. The face shovel driver in an earthworks team, which includes several dump trucks, a bulldozer and roller, is the person who sets the pace and is the obvious choice to receive the largest share.

It is often said that a site with a good bonus scheme is self-organising, but this overstates the situation. It is true that with a bonus scheme there will be plenty of pressure to overcome problems, whether these are within the team or outside it. The pressures from the operatives are primarily to maintain the levels of bonus, not necessarily for maintaining production at the required level. A lowering of the target, or the allocation of some hours against a heading for which bonus is paid pro-rata to the earned bonus, will maintain a high level of bonus, but will not produce the correct saving to the Contractor.

Another problem is one of executing all the easy work first, leaving additional costs later in the programme. Take as an example the earthworks team of tractors and scrapers, bulldozers and compaction plant which is excavating in a cutting and placing the material in an embankment—as indicated in Figs. 1(a) and (b). The diagrams illustrate two methods of operation.

In Fig. 1(a) the haul distances and therefore the output varies while in Fig. 1(b) the haul distances are maintained at the average throughout the operation. In both cases the overall average haul will be the same. The method depicted in Fig. 1(b) should result in a higher efficiency, because the spreading, levelling and compaction effort required is constant. With the varying output method, Fig. 1(a), if the compaction team is kept the same, there is the danger of under-compaction of the fill when the haul distance is at its shortest, and the operatives will be claiming excessive bonuses.

Another situation that could arise with the earth moving team is that too much material could be left to be trimmed off the cutting slopes, or the embankment slopes could be over-filled. It is important to become aware of any of these problems before any overpayment of bonus is made, and that steps are taken to prevent them.

To create incentive to work, giving maximum benefit to both operatives and management, requires detailed short-term planning to ensure that sufficient work is available and that no problems such as insufficient setting-out, or materials of suitable quality and sufficient quantity, arise.

Plant

Constructional Plant is defined in clause 1(1)(c) as

> '... all appliances or things of whatsoever nature required in or about the construction completion and maintenance of the Works but does not include materials or other things intended to form part of the Permanent Works.'

This definition encompasses both mechanical and non-mechanical plant, to be used by the Contractor or his sub-contractors.

clause 14(3)

Clause 14(3) requires the Contractor, if requested by the Engineer, to submit details of the Constructional Plant to be used on Site. This is often associated with the stresses and strains put on the Permanent Works by the working methods to be adopted, but may be the basis for the Engineer's assessment of the viability of the Contractor's methods and programme.

clause 53

clause 63

Clause 53 is mainly concerned with protecting the Employer in the event that the Contractor has forfeiture forced on him, under clause 63, but it applies to all contracts, although in many cases little thought is given to it. The clause applies to Plant which includes Temporary Works and Materials as well as Constructional Plant, but does not apply to transport to and from the Site. The points affecting the normal working of the Contract are as follows.

(*a*) All plant owned by the Contractor, or a company in which he has a controlling interest, is deemed to be the property of the Employer while it is on Sitc.
(*b*) All agreements to hire plant must contain a condition allowing the employer to take over the hire agreement in case of forfeiture of the Contract. This would include hire agreements with a plant hire company, owned by the parent company of the Contractor.
(*c*) No Plant, except hired Plant, can be removed from the Site without the written consent of the Engineer, but consent should not be unreasonably withheld where the items are no longer immediately required on Site.
(*d*) If the Contractor fails to remove any of his plant, after Completion, within the reasonable time allowed by the Engineer, the Employer may sell it and deduct the expenses of so doing from the proceeds, paying the remainder to the Contractor.

Boreholes and exploratory excavation

clause 18

Clause 18 provides for the Engineer to give the Contractor an instruction, in writing, whenever he requires boreholes or exploratory excavation to be carried out. Such instruction is to be treated as a variation, unless there is a Provisional Sum or Prime Cost Item in the Bill of Quantities for the work.

There is no problem if, when the Contractor considers that boreholes and so on are required, the Engineer gives the instruction. Often the Engineer argues that it is not he, but the Contractor, who requires the boreholes and refuses to give the instruction. As a general rule the Engineer should give an instruction if the exploration is required to obtain some information affecting the design—e.g. locating services to determine the position for a new manhole; to find the extent of unsuitable material under a proposed foundation; or to bore ahead of a tunnel face to determine if ground treatment is necessary. The Engineer might also decide that exploratory excavations or boreholes are needed, if unanticipated physical conditions or artificial obstructions are encounterd.

The specification or the preamble to the Bill of Quantities usually has a section stressing the Contractor's obligation to locate all services and to protect them. The services may cross the line of the new excavation or run along it, but sometimes may be in the way of the new work. If this is the case, a variation to divert the service or change line or level of the new work is needed. The Contractor might well consider that any exploration done by him to locate such services would merit the issue of an instruction, under clause 18.

clause 18

Safety

clause 15(2)

clause 19(1)

clause 19(2)

clause 19(1)

Clause 15(2) states that the Contractor or his authorised agent or his representative shall be responsible for the safety of all operations. Clause 19(1) requires the Contractor to have full regard for the safety of all persons entitled to be on the Site, so far as it is within his control, and to keep the Site in an orderly state, appropriate to the avoidance of danger. Clause 19(2) puts a similar responsibility on the Employer or his other contractors in respect of any work done by them on the Site. Clause 19(1) also makes the Contractor responsible, at his own cost, for the provision of lights, guards, fencing, warning signs and watching in connection with the Works, as required by the Engineer or a competent statutory, or other, authority. Whereas there is no mention of this provision with regard to the Employer or his other contractors, the reference to 'in connection with the Works' does relieve the Contractor from providing these services in respect of other work. Under clause 8(2) the Contractor is fully responsible for

clause 8(2)

> 'the adequacy stability and safety of all Site operations and methods of construction'

except design and specification of the permanent and temporary Works, carried out by the Engineer.

Under the law the Contractor is required to comply with various regulations regarding safety and welfare including: the provision of safety clothing such as hard hats, eye and ear protection and dust filters; the keeping of records of specified regular inspection of such items as scaffolding, excavation supports, lifting machines and hoists, the recording of, and the reporting of accidents and dangerous incidents that occur in connection with the work; and many others.

The Site agent's duty, as the senior contractor's representative on the Site, is to ensure that all regulations are complied with, either by personally doing so, or by delegating certain tasks to his staff. The Site agent will need to consult the Company's safety officer to ensure that he has up-to-date knowledge of all requirements.

3. Time and extension of time

The clauses to which reference is made in this chapter are: 2(3), 7(1), 7(3), 12(1) and (3), 13(3), 14, 20(1), 21, 27(6), 31(1) and (2), 40(1), 41, 42(1), 43, 44, 46, 47(b), 48, 49, 51(1), 52(4) (b), 59B(4), 60(5) and (6), 61, 63, and Form of Tender (Appendix).

Time is one of the most important aspects of a construction contract and it is not surprising that the ICE *Conditions* contain a number of clauses dealing with one or other aspect of it. Briefly the Contractor is required to commence the Works

> 'on or as soon as as is reasonably possible after the Date for Commencement of the Works to be notified by the Engineer in writing' (clause 41)

clause 41

and to complete any Sections and the whole of the Works within the periods stated in the Form of Tender (Appendix) (clause 43).

clause 43

The Contract allows for the awards of extensions to the construction periods if certain events, listed in the ICE *Conditions*, occur. These events are either deemed to be within the control of the Employer or, at least, outside the control of both Contractor and Employer. If the Contractor does not complete the Works, or sections thereof, in construction or extended times allowed, the Employer is entitled to compensation by way of Liquidated Damages.

The ICE *Conditions* require claims for extension of time and monetary claims, arising from delays, to be notified by the Contractor and dealt with, by the Engineer, as separate issues. This chapter is confined to consideration of time and extension of time, while monetary claims are included in chapter 6.

Commencement

The Date for Commencement notified by the Engineer 'shall be within a reasonable time after the date of acceptance of the Tender' (clause 41) and after commencement 'the Contractor shall proceed with the Works with due expedition and without delay in accordance with the Contract' (clause 41). It can be important to the Contractor to know, at the time of Tender, the

clause 41

clause 41

actual calendar date, within limits, of the Date for Commencement, as this may have a considerable effect on the prices in his Tender. This applies particularly to the smaller job or to any project with a short, very tight, construction period, where the season in which the work takes place needs to be known. There are other reasons for wishing to know the date, such as the probable effect on the price of a forthcoming wages increase, or the need to place an order to meet a rolling date for steel piles.

Control by Contractor

The Contractor can exercise some control over the lateness of the date for Acceptance, albeit this will be of a rather negative nature. He can incorporate into his Tender a time limit, after the expiry of which the Tender can no longer be accepted without the Contractor's approval. A normal time limit to allow is two months, but this may be too much for a short construction period. If the Tender is accepted just before the expiry of the limit, the Contractor is still faced with the uncertainty of the lapse of time between Acceptance and Date for Commencement. As clause 41 states, the time between the two should be a reasonable time. If it were to be longer than the Contractor anticipated, he would be able to claim that it was not in accordance with clause 41. However, making a claim at the very beginning of a project may not augur well for its future success.

clause 41

As an alternative to inserting a time limit into the Tender, the Contractor could insert a proviso that his prices would apply only if the date of commencement was to be before a stated date. It is virtually certain that this would be taken to be a qualification of the Tender, which would only be considered if he also submitted an unqualified Tender. The prices in this unqualified offer would apply whatever the Date for Commencement, but would be subject to adjustment if this date did not satisfy the Contract Conditions, e.g. the time between Acceptance and Commencement was not reasonable. The Employer or Engineer could avoid some of the difficulty if the problems could be foreseen and the Tender documents were prepared to accommodate them. One measure that is sometimes taken to avoid problems is to place an order with a supplier for equipment or material—needed early in the project—which has to be confirmed and taken over by the successful Tenderer, e.g. under a Nominated Sub-contract. If this were not done the Contractor might argue that 'reasonable time' meant time long enough, after acceptance, to place the order and allow delivery at the early date required.

The Contract does not make a start on the Date for Commencement mandatory, but the Contractor must commence 'on or as soon as is reasonably possible after'. Nevertheless there should be a few days, or perhaps a week, between Acceptance and the Date for Commencement, to give the Contractor the opportunity to mobilise his resources without taking this time out of the Time for Completion. This is not significant with jobs that last years, but could make a tight time almost impossible for very small projects.

Time for completion

The Time for Completion, often referred to as the construction period, is the period, starting on the Date for Commencement, in which the Contractor is required to construct the Works to a state of substantial completion, where they can be used by the Employer (clauses 41, 43 and 48). For most civil engineering projects the period is fixed by the Employer and or Engineer during the compilation of the Tender Documents. However, in the recent past there have been occasions when the time has been left to the tenderers, and has had to be taken into account in assessment of the competitiveness of each Tender submitted.

clause 41
clause 43
clause 48

The ICE *Conditions* envisage the incorporation of more than one Time for Completion, because the term applies not only to the period for the Whole of the Works, but also to sections thereof. The times are inserted in the Form of Tender (Appendix). All the Times for Completion in the Contract are periods starting on the Date for Commencement (clause 43) although it is not uncommon for the start of some sections to be dependent on the prior completion of other sections, or even the completion of work by other contractors. This means that the section periods in the Appendix could be misleading, at first glance. The most common situation is where completion of some new installation is necessary before existing factory plant can be demolished to make way for more new construction. Where some restriction exists that affects the start of a section, this should be stated and detailed in the Contract Specification, unless this is absolutely obvious from the logic of construction, as would be the case if the foundations and superstructure of the same building were made separate sections. The main use of section periods is to ensure that the Contractor has a contractual obligation to complete, in a time which will cause no delay to subsequent work by the Employer or his other contractors. Civil contracts for oil refineries or chemical process plants, requiring the Contractor to construct foundations, on which others build the plants, could contain a dozen or more sectional completion requirements.

Form of
Tender
(Appendix)

clause 43

The inclusion of Times for Completion in the Contract does

not prevent the Contractor from completing the various phases of construction in shorter periods, if he is able to do so. The *ICE Conditions* also require the Employer to take over the Works or sections thereof when the Engineer considers them to be substantially complete. The Engineer is not entitled to wait until the expiry of any Time for Completion before fixing the date for substantial completion, unless there is a specific reference to such an entitlement in an amendment to the standard conditions.

The previous paragraph sets out the view held by Contractors, which Engineers and Employers do not always accept readily. Since the point arises so often the grounds for the contentions are set down in the following. The Contractor's obligation to complete, stated in clause 43, is

clause 43

> 'The whole of the Works and any Section required to be completed *within a particular time as stated in the Appendix to the Form of Tender shall be completed within the time so stated* or such extended time as may be allowed under clause 44) calculated from the Date for Commencement of the Works notified under clause 41.'

My italics emphasise the words relied on to show that completion is not required on a date, but within a period of time. The times stated in the Appendix to the Form of Tender are Times for Completion. The use of the words 'within the prescribed time or any extension thereof' in clause 47(1)(b) in the context of what happens if the Contractor fails to meet his obligations with regard to time confirms the conclusion drawn from clause 43. In accordance with clause 48(1)

clause 47(1) (b)

clause 43

clause 48(1)

> '*When the Contractor shall consider that the whole of the Works has been substantially completed and has satisfactorily passed any final test* that may be prescribed by the Contract *he may give a notice to that effect to the Engineer* or to the Engineer's Representative accompanied by an undertaking to finish any outstanding work during the Period of Maintenance. *Such notice* and undertaking *will be in writing and shall be deemed to be a request from the Contractor for the Engineer to issue a Certificate of Completion* in respect of the Works and *the Engineer shall* within 21 days of the date of delivery of such notice *either issue to the Contractor* (with a copy to the Employer) *a Certificate of Completion* stating the date on which in his opinion the Works were substantially completed in accordance with the Contract *or else give instructions* in writing *to the Contractor specifying all the work which in the Engineer's opinion requires to be done by the Contractor before the issue of such certificate . . .*'

The italics show the important elements of the ICE *Conditions* under which a Certificate of Completion has to be issued by the Engineer. The Engineer has to issue the Certificate when he considers that the Works have been substantially completed, given that there is no proviso that allows the Certificate to be withheld until the expiry of the Time for Completion.

The ICE *Conditions* may well allow the Contractor to finish early, but there can be a number of practical difficulties which strain relations if the amount of reduction is very great. The reduction of time taken to complete a project would produce an amendment reduction in the recurring costs of running the site organisation, or so the theory goes. In the early 1970s some contractors appeared to take up the theory in a big way, and cut tender prices on the assumption that large reductions in the construction time could be achieved. In some cases, the assumption was correct, but whereas Contractors tended to blame delays and variations for their failure to achieve a reduction, the Employer and Engineer were of the opinion that these were only partly responsible. Proposed large time savings were looked on with considerable scepticism by Engineers. Some of this scepticism was no doubt prompted by the knowledge that, if the Contractor planned to complete in half the Contract Time for Completion, his preliminaries would be spread over that half of the time only, making any claim for delay approximately twice as large as would otherwise be the case.

Quite apart from the concern regarding the size of future claims, a proposal to complete in a fraction of the Contract time could pose problems for the Engineer and Employer such as the following.

- Early Completion would be of no benefit if the Employer was unable to use the project. This would apply, for example, if use depended on the completion of other associated projects.
- The Employer could have a problem in changing his budget, to make enough money available to pay the total value in the shorter time.
- The Employer might have difficulty in producing materials, which he had undertaken to supply.
- The Engineer might be unable to produce drawings fast enough for the quicker schedule.
- Other contractors, on whom completion was dependent, might also have to speed up to avoid causing delay. Of course, if the times for these were included in the Contract, the Contractor would be aware that his shorter construction period would not be viable.

If some of these disadvantages cannot be resolved, then the Contractor's and Employer's requirements with regard to the time for completion will be at odds with one another. For instance if the Employer has a problem with a budget, the Contractor may consider that it is no concern of his, but he is likely to experience delays in being paid.

The Contractor could have put in an alternative tender, containing his shorter Time for Completion, which if accepted would avoid any differences. As has already been said in chapter 1, contractors are wary of giving alternative bids which might give away some advantage to other tenderers, if they were given the opportunity to make an offer on the same basis. Having obtained a contract on one time, the Contractor, after acceptance, could try to negotiate a shorter time with the Employer and include this amendment in the signed Contract. In such a case the Contractor would be subject to damages if he were to be in default on the time, but at least it should convince both Engineer and Employer that the change in time was in good faith, and not a strategy to enhance claims.

Where the Contractor suggests the incorporation of a shorter time in the Contract, in order to avoid the complication sometimes associated with a programme much shorter than the Contract period, he would not expect to reduce his prices. However, if the Employer was required to amend some other provision, to make early completion possible, the Contractor might be expected to offer some reduction in price. If the Employer is in a position to gain from the shorter time, the Contractor may be able to obtain some increase in certain rates to compensate him for running a higher risk of incurring damages for delay. The advantage may not necessarily be one of using the project earlier, but could just give a larger safety factor for avoiding delays to following work, where this was much larger than the civil Works. The Contractor should give consideration to these points, but if necessary he might decide just to go ahead with a shorter time. The decision regarding amendments to a contract would lie within the Contractor's head office, not with the Site agent, although he may have been responsible for the idea.

Where the Contractor is required to incorporate his own Times for Completion into his Tender, and that Tender is accepted, the Contract will then be operated in the same way as if the Employer had given the Times.

The Department of Transport has been experimenting with a number of different variants to the ICE *Conditions*, which place a higher emphasis on the importance of Time. One of these does not involve any time being included in the Contract.

The Contractor is required to incorporate a sum into his Tender which represents his estimate of time he will occupy a carriageway, multiplied by the rent per day, for that carriageway, supplied by the Employer. Whatever the Contractor's estimate of time, he pays back to the Employer, by deductions from interim certificates, the actual number of days of occupation times the rent per day. No payment is due to the Employer in respect of any extra occupation days caused by variations or any of the other events that can lead to an extension of time under the ICE *Conditions*. The emphasis on time comes not only from the competition between contractors with regard to the period of occupation, but also because rents for a carriageway have been at levels two or three times those of the Liquidated Damages previously put into this type of contract.

The time taken by a Contractor to complete a project is subject to the effects of random circumstances, such as weather, human error, labour situation, materials deliveries, plant breakdowns and accidents, which can either extend or shorten the average time. The efficiency of the management team can lessen some of the effects, but there will always be some range within which times will vary. If a Contractor is to be sure, or 95% sure, of finishing the work in time then he needs to target some date before completion, so that virtually the whole range caused by random effects lies within the contract period. The saving in time is only likely to be a maximum 5–10% for this reason, but the statistical principle should be recognised by all involved.

Programme

It is unlikely that the Contractor will be given a completely free hand in drawing up a programme. The Contract Documents are likely to contain some constraints which must be built into the programme, such as

- section completions
- operations of other contractors
- completion of accommodation works before starting others
- order of release of land
- maximum rate of rise of embankments
- amount of money available for interim payments
- order of release of design details
- effect of method of construction on the Works as designed
- limitation of use of certain roads.

On some more complex projects the Contractor will be given a copy of the overall project programme, within which he must plan his own work. General references are made in the ICE

Conditions to the possibility of such restraints applying e.g.

clause 7(1)
clause 14(4)
clause 31(1)
clause 40(1)
clause 42(1)
clause 51(1)
Form of Tender (Appendix)

- clause 7(1), further drawings
- clause 14(4), effect of method of construction
- clause 31(1), other contractors
- clause 40(1), suspensions
- clause 42(1), order in which the work is to be executed and possession of the relevant parts of the Site
- clause 51(1), specified sequence method or timing of construction
- Appendix to the Form of Tender, Sectional Completions.

clause 43

However, full details of any constraints, known by the Engineer or Employer, should be given elsewhere in the tender documents, and the Specification is the most appropriate place. For instance, where section completions are incorporated into the Contract, the construction periods, given for the sections in the Appendix, all start on the Date for Commencement (clause 43). However, the work in a particular section may not be able to be started until some other section is completed, or until some preparatory work has been carried out by another contractor. It is reasonable for the Contractor to assume that all of the sections can be started on the Date for Commencement, in the absence of anything to the contrary in any of the documents that form the Contract.

clause 31(2)

However, the absence of specific, written details of restraints in the documents does not allow the Contractor to assume, always, that none exists. Some conditions require the Contractor to use his skill and experience to decide what should be allowed in his programme. Clause 31(2) is a case in point, as it gives the Contractor the right to extra time, because of delay by another contractor, only where that delay is

> 'beyond that reasonably to be foreseen by an experienced contractor at the time of tender'.

clause 14(6)

Similar provisions are contained in other clauses—e.g. clause 14(6)—in respect of delays that result from the Contractor's method of working's having to be altered because the method originally proposed adversely affected the permanent Works.

clause 14(1)

In accordance with clause 14(1):

> 'Within 21 days after the acceptance of his Tender the Contractor shall submit to the Engineer for his approval a programme showing the order of procedure in which he proposes to carry out the Works and thereafter furnish such further details and information as the Engineer may reasonably require in regard thereto.'

Owing to shortage of time, the programme submitted within 21 days is likely to be on a fairly broad basis, hence the provision for the submission of further details, as required. At the same time as he submits his programme, the Contractor is required to provide a general description of his proposed arrangements and methods of construction for executing the Works and, thereafter, furnish such further details as the Engineer may reasonably request, including calculations for the stresses and deflections imposed on the permanent Works during construction. Sometimes the Contractor is required to submit his proposed programme with his tender, and the proposals then become one of the contract documents. Such a programme would be an outline, showing significant items and times only, but the Contractor will be in default of his Contract obligations if he does not keep to the programme.

The ICE *Conditions* have no specific requirements of the details to be included on a programme and, to some extent, these could be expected to depend on the personalities involved. However, if the programme is to be of any use at a later date certain minimum information is necessary. Clause 14(2) entitles the Engineer to require the Contractor to produce a revised programme, if actual progress does not conform to that shown on the original programme. The proper comparison of programme and progress is extremely difficult unless the programme shows sufficient details of: location of activities; separation of preliminary temporary work items from permanent work items; and the breaking of a large operation into a number of smaller ones, if the speed of working varies.

clause 14(2)

The operation of the extension of time provisions also requires a detailed programme to verify the effect of delays on progress, by the Engineer or Employer, e.g. clause 42 entitles the Contractor to possession of parts of the Site to enable him to carry out work in accordance with his programme. This implies that the location of operations will be indicated on the programme. It is fairly normal now to give latest dates for such items as: receipt of drawings, ordering of materials and precasting off site; as part of the clause 14 programme.

clause 42

clause 14

Approval of the Contractor's programme

In accordance with clause 14(1) the programme is submitted to the Engineer for his approval, which the Contractor would not expect to be unreasonably withheld. Some consideration has already been given to possible objections, if the full Time for Completion is shown not to be required, or that insufficient details were given. Other possible criticisms could be as follows.

- The type of programme was not acceptable. This, however, would not be justified if the Contract contained no requirement for the format to be used. In such circumstances the Engineer would have missed his opportunity to specify the type of programme—in the Tender documents—but could give an instruction, under clause 13(1).
- The rates of output anticipated were not considered to be attainable. There appears to be no reason for the Engineer to do other than just make his point. If the rates are not attainable, then progress will be seen not to match programme, enabling the Engineer to request a revised programme (clause 14(2)).
- The Contractor had not built in all the constraints included in the Contract. This would be settled based on the facts, but might be contentious if the criticism was based on what an experienced contractor should have foreseen.

clause 13(1)

clause 14(2)

Unless the Contract contains some specific provisions, much of the information shown on a programme—such as choice of plant, areas of working and methods of construction—are within the Contractor's discretion. The Engineer should not refuse to approve a programme on a difference of opinion with regard to those matters. Neither, however, should the Site agent become involved in niggling arguments about the programme with the Engineer or his Representative, or work on the basis of one programme for the Contractor and one for the Engineer, except where the two show the same information in different formats.

Often the Contractor prepares and considers a number of programmes before settling on the one to be submitted. If each of these is uniquely numbered, and the number of the programme submitted stated in the covering letter, the original programme can be identified at any time in the future.

Programme revisions

As has already been mentioned, the Engineer is entitled to require the Contractor to submit a revised programme, showing the modification to the original to ensure completions within the Times for Completion, or extended times, if progress does not conform to programme (clause 14(4)). Under this entitlement the Engineer could require an up-dated programme if the progress was behind or ahead of programme, or if the Contractor had changed his original sequence of operations, or each time the Contractor was awarded an extension

clause 14(4)

clause 14(4)

of time, because in all cases progress would not conform to programme. There is no mention in clause 14(4) that the revised programme has to be approved by the Engineer, but it would need to accommodate all Contract requirements and constraints.

clause 46

Clause 46, although it does not specifically mention programmes, is also relevant. It deals with the situation where, in the Engineer's opinion, the Contractor's progress is too slow to ensure completion within time or extended time, but there is no further extension of time entitlement. In these circumstances the Engineer can notify the Contractor of his opinion and the Contractor has to take such steps as may be necessary and which the Engineer may approve, to expedite progress to complete in time. This will involve a revised programme, under clause 14(4). The Engineer is given no authority to tell the Contractor what steps to take to expedite progress, but only to approve or disapprove the Contractor's proposals. However, in serious cases of failure to take effective action, the Engineer might consider advising the Employer to invoke clause 63, or at least remind the Contractor that such a course of action is possible (chapter 7).

clause 14(4)

clause 63

Programme formats

The more usual types of programme used on civil engineering projects are

- bar chart
- network analysis
- time–chainage.

Each type of programme can be effective if used in the appropriate circumstances and prepared in sufficient detail, after due consideration, with calculations where necessary.

Bar charts. The bar chart is the simplest programme and, because one can be drawn without too much thought, is sometimes so short of detail as to be of little use in understanding the real state of progress made, or the steps that need to be taken to ensure timely completion. The worst bar charts give no indication of location, or the interdependence of the various activities, and nothing but overall rate of working. However, bar charts can be produced that remedy these faults, by splitting up items to indicate location, showing lead times for sequential dependent items and indicating resources for each item. The resource teams and rates of working should be given for each part of a conglomerate item, with a different output. Perhaps the prime example is the placing of the excavation and

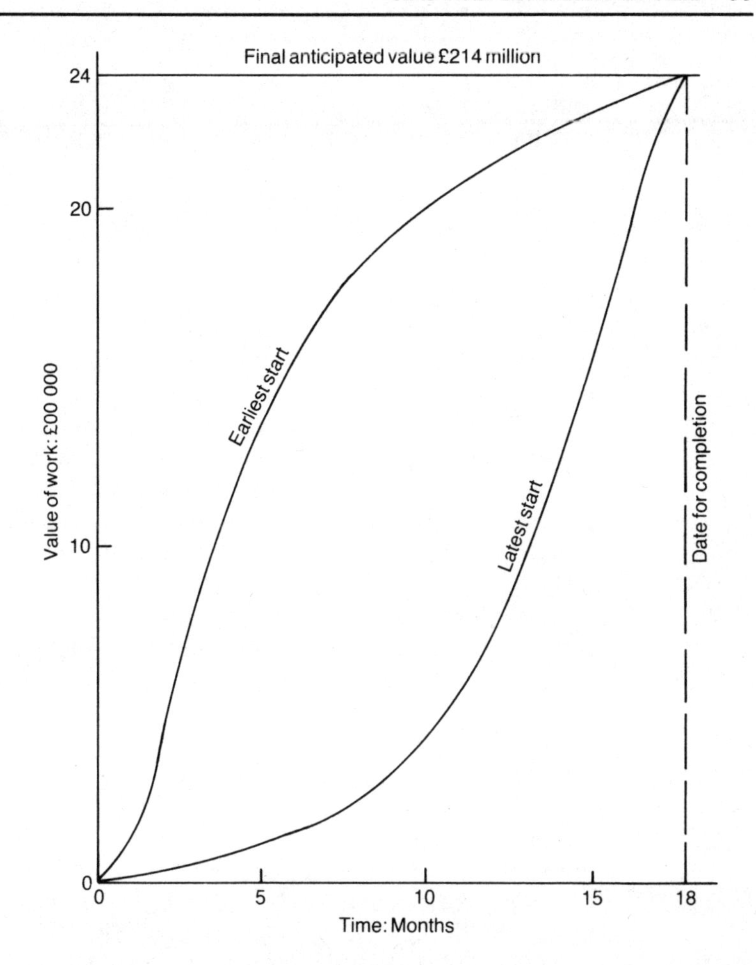

Fig. 2. Value of work against time for earliest and latest starts

filling of earthworks, in a number of materials and with a large range of haul distances, under one Bill of Quantities item.

Network analysis programmes. Timed network analyses programmes are usually considered superior to bar charts, from an engineering point of view, because of the calculations necessary for their production. Simple projects do not need the use of a computer for making up a programme and a bar chart would probably do as well. Complex projects definitely do need a computer, if the programme is to be kept up-to-date, and this opens up the possibility for other tasks such as resource levelling and programme valuations. It is not wise to assume, however, that the use of a computer ensures the output of a good programme. Computers are as adept at processing rubbish as they are in making calculations from sound data, and a network analysis programme, like any other format, is only as good as the logic and other data used in its production.

A basic network shows the order in which the activities take place and the restraints between items. Timing for each activity can be added, which allows the length of the various sequential paths to completion to be calculated. The longest or critical path and the earliest and latest start times for the non-critical activities can be scheduled from these calculations. Resources can be applied to all of the operations shown on the programme, and the non-critical items adjusted, in time, within the earliest and latest start dates, in order to level out resources throughout the construction period. Much is made of the importance of the controlling effect on completion of the critical items. If progress on critical path activities is on programme, there is a tendency to assume that the non-critical items will look after themselves. This can lead to a shock later on in the programme, when the importance of possible limitation of resources may come to light.

The problem is illustrated by Fig. 2, which shows the two cumulative values of work to be done, against time, for a programme, assuming all work is started on the earliest and latest start dates. The resulting graphs form an envelope, of which the lower boundary represents the latest and the upper one the earliest start dates. The width of the envelope depends on the amount of work that is non-critical, and the average amount of float between the earliest and latest start dates. An envelope reduced to a single line would represent a programme in which all the activities were critical. The slope of the graphs at any point represents the rate of working at that time.

The rate varies considerably throughout the period, with the greatest rates being in the earlier days for the earliest start line, and in the closing stages for the latest start line. In both cases the highest output required to complete on the last day of the contract period is more than double the average output required.

The envelope drawn is very wide, and after seven and a half months the critical items could be on programme, with as little as £200 000 of work completed, but with an ever-increasing output per month required to maintain programme. An increasing output does not necessarily involve an increase in labour and plant resources of the same magnitude, if, for instance, large bought-in items are to be installed late in the programme. A truer picture would be obtained by drawing particular resources against time, instead of output, e.g. number of men required, each week, for earliest and latest start times. In general, it is both easier to maintain programme and cheaper in cost to use average resources rather than fluctuating ones, but restraints on rate of working, such as winter weather, have to be taken into account.

Time–chainage programmes. Time–chainage programmes are suitable for the linear type of job such as sewers, water mains, roads, sea walls or even tall chimneys, where logic is fairly simple and location is the real key to the programme. Often the time–chainage programme needs to be augmented by other types of programming for parts of the work which are not linear, such as a pumping station on a main sewer contract or bridges and roundabouts on road contracts. The big advantage of these programmes is that location of operations, the direction of working, and the movements, such as those between cutting and filling, can all be shown. Fig. 3 illustrates a part of a time–chainage programme for a road project.

Extension of time

Form of Tender (Appendix)

The Time for Completion for Sections and the whole of the Works, given in the Form of Tender (Appendix), are subject to extension should certain events mentioned in the ICE *Conditions* occur and cause prolongation beyond the then current completion date. The events fall into two broad categories

- Events deemed to be within the control of the Employer, the Engineer or the Employer's other contractors or workmen
- events deemed to be outside the control of both Contractor and Employer.

clause 12

The general significance of the two categories is that, for the first category events, the Contractor is entitled not only to the extension but also to reimbursement for any expense caused by the event; second category events give a right to an extension of time only. A Contract can incorporate the Contractor's right to payment for a second category event, and this is the case in respect of delays due to unforeseen physical conditions or artificial obstructions (clause 12). In any Contract, the Contractor will only be entitled to payment for expense caused by an event if the conditions contain a clause that provides for that recovery.

The Contractor is also entitled to have the Time for Completion adjusted if the Employer's breach of Contract causes a delay in completion. A possible example is the delay caused by the Employer's failure to supply some materials, at the time he had undertaken to do so in the Contract.

Many people have the impression that the award of an

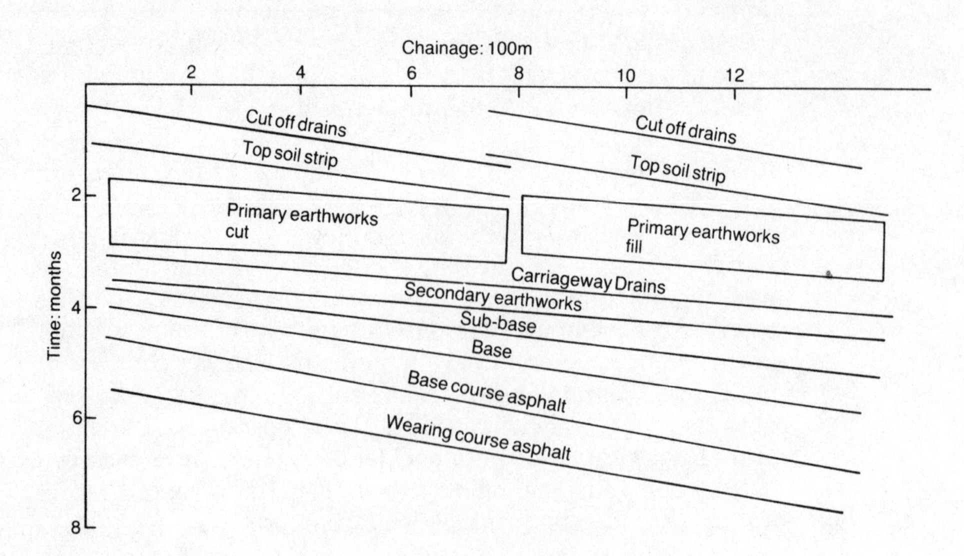

Fig. 3. Part of time–chainage programme for road contract

extension of time and reimbursement for delay are intrinsically linked, but this is by no means correct. Of the delays that give rise to extensions, mentioned in clause 44(1), and listed in detail in Appendix 3, two are not covered by any other clause which allows the Contractor to claim for payment. These are those due to 'exceptional adverse weather conditions or other special circumstances of any kind whatsoever . . .'. It is also true that an event that causes a delay, can entitle the Contractor to reimbursement, either on a cost or amended rates basis, but not to an extension of time. This is illustrated in the following example. (Whether a claim is evaluated on cost or on amended rates depends on the cause of the delay, and is covered in chapter 6.)

clause 44(1)

Figure 4 depicts the overall programme of a project, consisting of building A and building B. If there is a sectional completion date for building A, and a delay of six weeks in its construction occurs, owing to late issue of drawings, the Contractor would be entitled to both cost and an extension of time, for the section (clauses 7(3), 44 and 52(4b)). The Contractor would have no claim for extension for the whole of the Works unless he could show that the completion of the construction of building B depended on the release of resources from building A.

clause 7(3)
clause 44
clause 52(4b)

In contrast, consider the same situation and events, except that the Contract contained no Sectional Completion Time for building B. Unless late release of resources from building A affected building B, there would be no overrun of any Contract

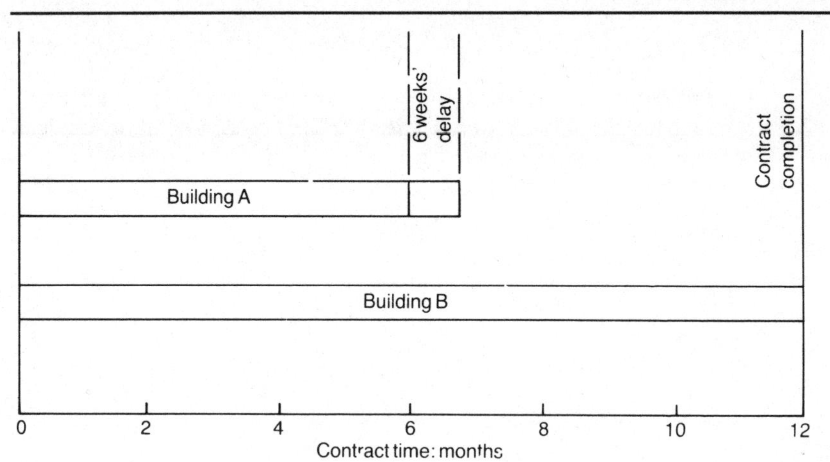

Fig. 4. Basic bar chart programme for two structures

time, so no extension of time could be awarded. Nevertheless, the effect on the cost of construction of building A would be exactly the same as before, and the Contractor should be able to recover the additional cost, under clauses 7(3) and 52(4b). (clause 7(3), clause 52(4b))

The procedure for the Contractor to make a claim for an extension of time is contained in clause 44(1), and the procedure for the Engineer to award extensions is given in clauses 44(2)–(4). Three sub-clauses are necessary to cover awards of extensions, because the Engineer is required to give consideration at three times (clause 44(1), clause 44(2), clause 44(3), clause 44(4))

- when the cause for extension arises (clause 44(2)) (clause 44(2))
- if the current date for completion for a Section, or the whole of the Works, is passed, before completion is achieved (clause 44(3)) (clause 44(3))
- if the contractor is in apparent default on time, for a section or the whole of the Works, when the relevant certificate of completion is issued (clause 44(4)). (clause 44(4))

Table 1 indicates the interrelationship between these clauses and those that mention the events that give rise to an entitlement to an extension of time. Appendix 3 contains a summary of the clauses relevant to extension of time.

Contractor's claim for extension—clause 44(1) (clause 44(1))

The Contractor's claim must be addressed to the Engineer, because the Engineer's powers to award extensions of time cannot be delegated to the Engineer's Representative (clause 2(3)). However, it would be imprudent not to give the Resident Engineer a copy of all letters sent to the Engineer on this subject. (clause 2(3))

Table 1. Interrelationship between clauses pertinent to extension of time

Clauses giving events entitling the Contractor to an extension of time	7(3), 12(1), 12(3), 13(3), 14(5), 14(6), 27(6), 31(2), 40(1), 42(1), 44(1), 59B(4)
Clause giving the procedure for the Contractor to claim an extension of time	44(1)
Clauses giving the procedure for the Engineer in awarding an extension of time	44(2), 44(3), 44(4)

clause 44(1)

If any cause of delay occurs, which the Contractor believes entitles him to an extension of time for the completion of the Works or a Section thereof, clause 44(1) requires that the Contractor

> '. . . shall within 28 days after the cause of delay has arisen or as soon thereafter as is reasonable in all the circumstances deliver to the Engineer full and detailed particulars of any claim to extension of time to which he may consider himself entitled in order that such claim may be investigated at the time.'

The period of '28 days or as soon thereafter as is reasonable' in which the Contractor is required to submit his claim, commences when the cause of delay arises. In the case of a variation, the cause of delay could be argued to arise when the variation is issued, although the work may be done and the delay incurred at a much later time. The Contractor should inform the Engineer of any significant potential delay as soon as possible, although the full extent may not be known. This will avoid possible embarrassment for the Engineer, who may be unaware of the effect of the variation, and will avoid the Contractor having to spend time in countering such arguments as 'if I had only known of the delay at the time, I would have amended the variation'. For smaller delays it might be sensible to review the situation, perhaps every two months, and make one application to cover a number of these items.

At the time a delay arises there may seem to be no need for an extension, if for instance the delay is not on the critical path, or if, at that time, the programme is aimed at achieving an early completion. The delay should be noted, in case the assumptions fed in to the programme prove to be wrong, or the contingency time on the programme is required because of the

Contractor's delays. The Contractor's detailed application for an extension of time should demonstrate the effect of delays, and the vehicle for this is the programme submitted by him, in accordance with clause 14(1).

clause 14(1)

clause 42(1)

Clause 42(1) indicates that if the Contractor is not given possession of the Site, or parts thereof, in time to allow working in accordance with his programme, then the delay in possession has to be taken into account in determining the extension due. Taking the lead from this clause, it is reasonable to base on the programme all other delays that are to be taken into account in the assessment of the entitlement to extension. There are two types of effect on the planned sequence of construction which can prolong the anticipated completion,

- Events which cause delay to existing activities and thus cause delay to overall completion. The effect can be direct, if the delay is on a critical activity or a near-critical one made critical by the delay. The indirect effect applies where the full resources (available at the time of the delay) cannot be utilised, because of the delay, which leads to a need for a higher, unobtainable level of resources later.
- Additional work which requires extra time, either because the work is in the existing or new critical path, or because of a similar indirect effect to that above.

Some variations can cause both the need for time to carry out the extra work, and a delay. An example of this is a variation that affects the critical path activities, and involves the ordering of materials, which cannot be obtained in time to avoid delaying other critical path operations. The indirect effect mentioned is only indirect because all the restraints that cause criticality were not examined in formulation of the programme. To cause delay of this kind the overall level of availability of some categories of resources has to be critical.

In calculating the extension required, account should be taken of the season into which the work will be extended, where the final operations are particularly weather susceptible. If work were to be extended into the winter, from autumn, Contractors would claim any additional extension, due to prolongation in the winter, under the heading for the primary event that caused the initial delay. Claiming extensions of time due to delays caused by

> '... exceptional adverse weather conditions or other special circumstances of any kind whatsoever which may occur be such as fairly to entitle the Contractor to an extension of time ...' (clause 44(1))

clause 44(1)

is not popular with Contractors, because costs of the delays are not reimbursible, but extensions under these headings are preferable to payment of Liquidated Damages for delay.

The interpretation of 'exceptional adverse weather' varies considerably among Engineers, and can prove a problem in the establishment of a claim, what with the very variable weather which is normal in the UK. 'Exceptional' could be above average or anything within three standard deviations of the mean, or anything in between—depending on opinion, and mathematical bent. All worse than average weather, which occurs 50% of the time, cannot be exceptional. However, if the Time for Completion, inserted into contracts, were to be calculated to allow the Contractor to complete in time for any weather within three standard deviations of the mean, these times would be much longer than they are at present. Some practical compromise needs to be made, taking into account the weather that affects the susceptible items over the period in which they are programmed. For instance, if a tower crane is to be used for three months, it is the incidence of winds strong enough to prevent its operation in that period which is important and not the incidence over the full year.

Events coming within the term 'other circumstances of any kind whatsoever' might include the occurrence of strikes, materials or labour shortages not reasonably foreseeable at the time of Tender, accidental damage to the Temporary or Permanent Works, or vandalism despite the Contractor's having taken reasonable steps to protect the project.

If a claim is being made for decrease in productivity caused by constant delays and changes (chapter 6), this should be examined to see if it has an effect on time.

When a delay has occurred to one section of the project, a contemporary claim for extension to other sections or the whole of the Works should be made, if these are affected.

If a computer-run, network analysis programme is being used then the individual changes to activity times can be fed into the computer to give a revised completion time to be included in the notice of claim. With a really complex computer programme, it would be possible to do a run limiting resources to those required in a resource-levelled original programme. However, the use of the more complex techniques does not extend to all contracts, and there the proof of the effect of delays on completion tends to be much more approximate. Calculations such as the number of man-days required for additional work divided by the average labour force, or the amount of disruption caused (chapter 6) can be a basis for an estimate of the extension entitlement.

The Engineer's assessment and awards of extension
The Contract requires the Engineer to give consideration to the Contractor's entitlement to extensions of time at three stages

- on receipt of a claim from the Contractor or, if he thinks fit, in the absence of such a claim; but presumably at the time a claim should have been submitted (clause 44(2))
- as soon as possible after the passing of the then current contract completion date for a section, or the whole of the Works (clause 44(3))
- on the issue of a Certificate of Completion of the Works or the relevant Section (clause 44(4))

clause 44(2)

clause 44(3)

clause 44(4)

In all three cases the Engineer should make a definite decision, and either award the Contractor an extension or inform the Contractor, in writing, that no entitlement is considered to be outstanding. The decision should be made after the Engineer has considered and taken into account all the circumstances known to him at the time, not only those mentioned in the Contractor's notice.

clause 44(2) clause 44(3)

Under clauses 44(2) and (3) the Engineer is required to make

> 'an assessment of the extension of time (if any) to which he considers the Contractor entitled for the completion of the Works or relevant Section . . .'.

clause 44(4)

The requirement under clause 44(4) is that

> 'The Engineer shall upon the issue of the Certificate of Completion of the Works or of the relevant Section review all the circumstances of the kind referred to in sub-clause 1 of this Clause and shall finally determine the overall extension of time (if any) to which he considers the Contractor entitled in respect of the Works or any relevant Section. No such final review of the circumstances shall result in a decrease in any extension of time already granted by the Engineer pursuant to sub-clause 2 and 3 of this Clause.'

clause 44(2)

As a result of the use of the word 'assessment' in clause 44(2), and the fact that the final extension cannot be less than an interim extension, there is sometimes a tendency for the Engineer to be cautious on interim awards. The Contractor should object to any more obvious instances of awards which do not take the full circumstances, known at that time, into account.

A case of the need to review extensions at a later stage in the Contract, is where the delay, caused by a qualifying event, is less than the overall float shown on the programme. The Contractor should have noted the delay, and should raise it again, in the event that anticipated progress is not achieved, making an overrun probable. Another case is where the extension runs into winter, which may cause further delays to weather-susceptible work.

The delays taken into account are those which are 'such as fairly entitle the Contractor to an extension of time . . .'. The Engineer would make a decision that is fair to both Contractor and Employer.

Clause 47—Liquidated Damages and Limitation of Damages for delayed completion

clause 47

If the Contractor fails to complete the whole or a section of the Works within the relevant time period, as extended in accordance with the contract provisions for awards of extension of time, he is liable to pay the Employer damages for that breach of contract.

In the case of contracts based on the ICE *Conditions*, these damages are not the actual damages suffered by the Employer, as a result of the delay, but Liquidated Damages. These are, as described in clause 47(1) (a)

clause 47(1) (a)

> '. . . the sum which represents the Employer's genuine pre-estimate (expressed as a rate per week or per day as the case may be) of the damages likely to be suffered by him . . .'.

Liquidated Damages may turn out to be less or more than the actual damages incurred by the Employer. Any estimate can be overtaken by intervening events, between tender and completion, but the Employer is entitled to the pre-estimated amount, whether it is more or less than actual.

clause 47(1) (a)

Clause 47(1) (a) also provides the alternative of the substitution of a lesser sum, which represents the limit of the Contractor's liability for damages for failure to complete, in lieu of the genuine pre-estimate for damages. The amounts for Liquidated Damages should be inserted in the Form of Tender (Appendix) by the Employer, or the Engineer on his behalf, before the tenders are invited, but are very occasionally omitted.

Form of Tender (Appendix)

The Engineer will often be responsible for the calculation of the rates for damages, presumably on information supplied by the Employer. Such information is needed if the rates are to be a genuine pre-estimate, but many figures included in tenders appear to be based on the amount tied up in the Contract, or in the overall project. Rates seen in Tenders, for Liquidated Damages, have varied enormously between the equivalent of

7.5% and 87.5% of the Contract value for a year's delay in completion of the whole of the Works. The smaller figure can only be a 'lesser sum', not a genuine pre-estimate. The very high figure is likely to arise in cases where the Contract is part of a very much larger project. It would not be prudent for an Engineer to insert any rate other than a genuine pre-estimate, without the Employer's prior knowledge and agreement.

The type of items in a calculation for a Liquidated Damages rate might include

- additional costs over the period of the delay—e.g. Resident Engineering staff and accommodation, claims from following contractors equipment suppliers and so on, if they are affected by the delay
- costs connected with the intended use of the structure which cannot be avoided and are not offset by income from its use—e.g. staff employed, interest on money invested, ground rents; these would be calculated for the whole project, rather than for the civil works only, if the whole was affected by the late completion of any section of the civil works
- loss of profits on not being able to use the structure or project at all, or not as efficiently as when completed.

Where the Employer is not a profit-making organisation and there are no following operations (e.g. on Department of Transport road contracts) the Liquidated Damages can only be calculated on a notional basis, such as the interest on the amount tied up in the project, in addition to the prolongation costs for the Employer's supervision.

Where a number of operations, by the Employer or his other contractors, follow the Contractor's progress on his part of the overall project, there is need for a number of Liquidated Damages amounts to preserve the genuine pre-estimate basis. This arises because the amount of cost from delays to other operations must depend on where any delay occurs in the Contractor's part of the project. Delays early on could run right through, and affect all following contractors, but those at the end would affect some only. The ICE *Conditions* allow this to be done by the incorporation of Liquidated Damages for the whole of the Works and separate figures for each of the sections. It is difficult to visualise the results of the calculations for damages, where sectional damages are incorporated, without putting down the formulae for the various cases. These are illustrated in Appendix 4.

The examples, contained in Appendix 4, illustrate that strict use of the formula in clause 47 for calculating the Liquidated

clause 47

Damages, covering a number of possible events or combination of events, will not always result in a genuine pre-estimate of damages. Clause 47(3) states

clause 47(3)

> 'All sums payable by the Contractor to the Employer pursuant to this clause shall be paid as liquidated damages for delay and not as a penalty.'

If the sums are not genuine pre-estimates of the damage likely to occur, or are lesser sums, they can be challenged by the Contractor as being penalties. It is not the size of the rates for damages, alone, that is important, but the manner of their calculation.

It is accepted that the Site agent has no control over the figures for Liquidated Damages inserted in a Contract, but he should be aware of the significance of the rates. Knowledge of the calculation of genuine pre-estimates is helpful should the Liquidated Damages clause have been omitted from a particular contract and replaced by an actual damages clause. The Site agent's awareness of whether or not the rates for Liquidated Damages are lower than the actual damages likely to be incurred could be useful, should the possibility of negotiating an acceleration agreement arise.

The Employer is entitled to deduct any Liquidated Damages for delay from payments due to the Contractor. No deduction can be made until the Engineer has informed the Employer and Contractor, in writing, that he considers that the Contractor is not entitled to further extensions of time. This notification appears to be a necessary preliminary to the deduction of Liquidated Damages, even if the Contractor has made no claim for a further extension (clause 47(4)). If after the deduction of damages, the Engineer awards a further extension of time, then the Contractor is entitled to the reimbursement of any damages deducted, in respect of that period, together with interest at the rate(s) provided for in clause 60(6) (clause 47(5)).

clause 47(4)

clause 60(6)

clause 47(5)

Completion

clause 48

Clause 48 deals with substantial completion and the issue of Certificates of Completion in respect of a part, a section, or the whole, of the Works. A part of the Works has to be both substantially completed and also occupied by the Employer before a certificate can be issued, but a section or the whole of the Works needs only to be substantially completed. The procedure in all three cases is the same in so far as the following situations are concerned.

- The Contractor gives notice, in writing, to the Engineer or his Representative when he considers the relevant part, section or whole of the works has been substantially completed and passed any test required by the Contract. In the notice he gives an undertaking to complete any outstanding work during the Period of Maintenance. Clause 49(2) requires the Contractor to complete the outstanding work as soon as may be practicable after the date of completion.

clause 49(2)

- Within 21 days of delivery of the notice the Engineer must either issue the relevant certificate, giving the date of completion, with a copy to the Employer, or give the Contractor a schedule of the work that must be done before the certificate can be issued.
- Where a certificate is refused, at the first request, the Contractor is entitled to the certificate, within 21 days of completing the schedule of work, to the Engineer's satisfaction.

The clause allows the Contractor to give notice to either the Engineer or his Representative, but it should be remembered that authority under this clause cannot be delegated (clause 2(3)). It is the Engineer who must issue the Certificate of Completion, if it is to be valid. However, it is likely that the Engineer will rely very much on his Representative's report, in deciding whether or not to issue a certificate.

clause 2(3)

Although the Engineer has 21 days after receipt of the Contractor's notice to take action, the date for completion contained in the certificate is not the date of issue. If the Site agent and Engineer's Representative have previously agreed a list of items to be completed, and have made a joint inspection —before the notice had been given—the issue of the certificate should be assured. The action of ensuring that the agreed list of items needed for completion has been carried out allows the Site agent to state the date for completion, in his notice, and to be sure that it will be accepted. If it is not, the Contractor could be liable to Liquidated Damages during the 21 days. If the Site agent forgets to include an undertaking to complete outstanding work, during the Maintenance Period, in his notice, the Engineer is unable to issue a Completion Certificate (clause 47(1)).

clause 47(1)

Completion does not have to be absolute, but 'substantial', which is usually interpreted as in a condition which allows the construction to be used by the Employer, without serious inconvenience. 'Use' would include fitting out or commencement of the next phase of construction. The Site agent should be prepared for a tighter definition of 'substantial' if the Employer is unable to make use of the project, owing to some external restraint, compared with the situation where he is impatient to move in.

The issue of a Certificate of Completion gives a number of benefits to the Contractor:

- from the date of completion he is no longer liable to Liquidated Damages for the relevant part, section or whole of the Works (clause 47)
- he is entitled to be paid the part of the retention held for the relevant piece of the project, within 14 days of the issue (clause 60(5))
- his primary responsibility for care of the relevant piece of the Works ends 14 days after the issue, but he remains liable for any damage he causes, in the Maintenance Period (clause 20(1))
- the main insurance of the Works becomes the responsibility of the Employer 14 days after the issue of the Completion Certificate, but insurance to cover the Contractor's liability for damage he might cause in the Maintenance Period must be continued (clause 21)
- the Period of Maintenance begins for the relevant piece of the Works on the date of completion (clause 49(1)).

clause 47

clause 60(5)

clause 20(1)

clause 21

clause 49(1)

Maintenance certificate

The Period of Maintenance included in a civil engineering contract is inserted by the Employer or Engineer, in the Form of Tender (Appendix), but the most usual period is 52 weeks, commencing from the date of completion.

Form of Tender (Appendix)

There is only one Maintenance Certificate for a contract, despite the fact that a number of Certificates of Completion may have been issued (clause 61(1)). This certificate is issued by the Engineer to the Employer, with a copy to the Contractor, after the expiry of the latest period of maintenance, when

clause 61(1)

> '. . . all outstanding work referred to under clause 48 and all work of repair amendment reconstruction rectification and making good of defects, imperfections shrinkages and other faults referred to under clauses 49 and 50 shall have been completed . . .'.

The items of work, which must be completed before the certificate can be issued, include that which is required through no default of the Contractor, for which he will receive payment (clause 49(3)).

clause 49(3)

The issue of the Maintenance Certificate signifies that the Contractor has completed his obligations to construct complete and maintain the Works to the Engineer's satisfaction (clause 61(1)). However, the Employer's and Contractor's liability to each other, connected with their performance of

clause 61(1)

clause 61(2)

their contractual obligations, remains (clause 61(2)).

The payment of the second half of the retention money is not conditional on the issue of the Maintenance Certificate, but should be paid within 14 days of the expiry of the last Period of Maintenance. However, sufficient money can be withheld to cover the cost of any defects outstanding at that time (clause 60(5) (c)).

clause 60(5) (c)

4. Sureties, care and insurances

The clauses to which reference is made in this chapter are: 10, 18, 19, 20, 21, 22, 23, 24, 25, Form of Tender and (Appendix) and Form of Bond.

Sureties

clause 10

The provision for some kind of surety, under clause 10 of the ICE *Conditions*, does not appear to be a standard requirement, because the clause applies only if the Tender contains an undertaking, by the Contractor, to supply one when required.

Form of Tender

The undertaking made, if the Tender is on the Form of Tender, contained at the back of the ICE *Conditions* is

> 'If our Tender is accepted we will, when required, provide two good and sufficient sureties or obtain the guarantee of a Bank or Insurance Company (to be approved in either case by you) to be jointly and severally bound with us in a sum equal to the percentage of the Tender Total as defined in the said Conditions of Contract for the due performance of the Contract under the terms of a Bond in the form annexed to the Conditions of Contract.'

Despite the use of the words 'when required' rather than 'if required' Sureties or Bonds are not always required when the Form of Tender is used. If some form of guarantee is to be supplied it is normal to make a statement to this effect, somewhere in the Tender documents. The very least that is required is the insertion of the percentage of the Tender Total required, against the 'Amount of Bond' in the Form of Tender (Appendix).

Form of Tender (Appendix) clause 10

Where the provisions of clause 10 apply to a contract, then the Contractor undertakes to supply two sureties or, alternatively, a guarantee from an insurance company or bank, to the Employer. The two sureties mentioned in the clause refer to a practice from the past, in which friends or business acquaintances of the Contractor stood as surety that the Contractor would complete the Contract. Nowadays, the requirement is almost invariably one for the guarantee from an insurance company or a bank, in the form of a Bond.

A Bond is an agreement between three parties, two of whom

are involved in a separate contract agreement, and the third (the Surety or sometimes called the Bondsman) is joined with one of the parties to the contract to guarantee that party's performance of the contract, to the other. The guarantee is usually only to a limited amount, which, in the case of a Bond, under clause 10, should not exceed 10% of the Tender Total. How and when the Surety can be called on to pay up, under the Bond, depends on the terms of the Bond. However, under all Bonds, the Surety is entitled, by law, to recover from the Contractor any amount which he pays to the Employer. Therefore, the Bond tends to be a safeguard against the insolvency of the Contractor. It is not uncommon for the Surety to require that his common law right to recover, from the Contractor, any sums which the Surety has paid out to the Employer, should be backed up by a written agreement, making the Contractor's parent Company, if any, responsible for repayment.

clause 10

Normally the Contractor is responsible for paying the Surety's fee, for the risk he takes, and that amount is included in the tender price, but clause 10 does provide for other arrangements being stated in the Contract. The higher the risk, the larger the fee charged, until the situation is reached where the risk is considered so large that no Surety will issue a bond to that Contractor.

clause 10

Form of Bond

clause 10

Form of Tender

Although the ICE have produced a standard Form of Bond which is reproduced at the end of the ICE *Conditions*, the use of this form is not made mandatory by clause 10, which only refers to a bond, the terms of which are approved by the Employer. However, the use of the ICE Form of Bond is incorporated into the contractor's offer to carry out the project, where this is made on the ICE Form of Tender. In other cases, most contractors would feel justified in assuming that the ICE Form would be approved by the Employer, if no alternative form of bond was included in the Tender documentation. The ICE Form of Bond contains terms which most Contractors consider to be essential for any guarantee, as follows.

Form of Bond

Form of Bond

- There is a definite date when the Surety is discharged from his undertaking. In the case of the ICE Form, this is the date on the Maintenance Certificate (condition (c) on the Form of Bond).
- Payment by the Surety is made only on the default of the Contractor (condition (b) of the Form of Bond).

clause 10

Under clause 10, the Bond is to be from a Surety, and in terms approved by the Employer. If these are to be different from the ICE Form of Bond, it is only reasonable to give the terms in the tender documentation, because the terms could affect the fee for providing the Bond, or even a contractor's ability to obtain one on those terms.

The Bond is held by the Employer and, if no discharge date is included in the terms of the bond, the risk continues for the Surety, until the Bond can be retrieved from the Employer. In the case of the ICE Bond, the Surety is entitled to assume that the risk is continuing until he has proof that the Maintenance Certificate has been issued by the Engineer (condition (c) of the Bond).

Initially the Contractor pays a lump sum to the Surety, for undertaking the risk that there will be a call on the Bond, based on the assessment of the risk and the amount of the Bond. Additional payments could be involved, if there is a change in the risk.

Under the terms of the ICE Bond, the risk is run from the time of issue of the Bond until the issue of the Maintenance Certificate for the Whole of the Works. If the construction period is prolonged, either because there is an extension of the contract time or due to the Contractor's delay, then the risk will continue for a longer period than that anticipated when the Bond was issued.

The maximum to be paid out by the Surety on the Bond is a percentage of the Tender Total, which is unaffected by the final value of the Contract. However, if the Final Value of the Contract is significantly higher than the Tender Total, this could be regarded as a factor increasing the risk of a call on the Bond, and justification for an additional fee.

The terms of a Bond other than the ICE Bond should be included in the Tender Documentation, preferably by including a copy of the proposed form of bond. Terms that are put forward from time to time include

- a limit—which has been as high as 100% of the Tender Total
- omission of any end date or event, when the Surety will be released from his undertaking
- payment by the Surety is to be on the demand of the Employer, rather than on the default of the Contractor.

This last example of non-standard terms has been imported from Middle East contracts, and is removed from the conception of a bond as a protection against the insolvency of the Contractor, but can be used to exert unfair pressure on him. Quite apart from the fact that theoretically the bond can be called in at the whim of the Employer, it can also be called in the event of a dispute between the Employer and the Contractor regarding money, allegedly owed to the Employer. This would affect the Contractor's cash flow, and reduce his negotiating strength.

clause 10

There is some justification for a limit on the Bond in excess of the 10% maximum, included in clause 10, if the Contract is a small but extremely important part of a much larger project. In such a case the Liquidated Damages payable, on the default of the Contractor, could well reflect the value of the overall project, and be high in proportion to the Contract value. However, for such a contract, with enormous damages, it would be more prudent to reduce the risk when choosing the Contractors on the tender list. After all, the difference in quotes, as a percentage of the whole project, will only be a fraction of that as a percentage of the one civil construction contract.

Where there is no date or event to signify the end of the bond's cover, the risk to the Surety continues until the Contractor is able to retrieve the bond document from the Employer, and return it to the bondsman. The Contractor could reasonably argue that the need for a bond ends when the Certificate of Completion for the whole of the Works has been issued. At that stage, the work would have been executed to the satisfaction of the Engineer, and the one half of the retention, retained by the Employer, would be available to complete any maintenance, in the event that the Contractor is unable to do so. However, the Employer may opt to retain the bond after the issue of the Maintenance Certificate, forcing the Contractor to decide whether to grin and bear the additional cost or take some action to recover the bond. The dilemma should be avoided by the Contractor's refusal to accept a bond with no end date.

The Site agent would not normally be involved in the provision of the Bond, but could be involved with the Engineer, if he were dealing with the matter, in obtaining approval of a proposed Surety and terms for a Bond. Also, prompt issue of a bond is essential, particularly where the Contract requires the Bond to be sealed, before payment of any Certificate. Consideration also needs to be given to the possibility of Subcontractors supplying Bonds.

One of the items that should be included on the Site agent's list of items outstanding at the end of the Contract should be the provision, to the Surety, of the recovered Bond or other evidence that the risk of a call being made on the Bond has ceased.

Another form of protection that has become more popular with Employers, in the past few years, is the parent company guarantee, which can be a requirement of the Tender documents, quite separate from, and often in addition to, the bond. Contractors might lose some of their dislike of this requirement if those Employers who call for them did so to equalise the risk from all tenderers, but this is not always so. A subsidiary company can be required to give a parent company guarantee although the subsidiary may have a turnover ten or twenty times that of other tenderers on the list who have no parent. The Contractor should try to insist that any parent company guarantee will end on issue of the Certificate of Completion for the whole of the Works, but, like a bond, the guarantee should always contain a time limit.

The Site agent needs to ensure that any requirements for guarantees are met promptly, particularly where the Contract does not allow any work to commence or payments to be made to the Contractor until such arrangements are already in place.

Care and insurance of the Works

Care of the Works

clause 20(1)

Under clause 20(1) the Contractor is responsible for the care of any part of the Works from commencement up to 14 days after a Certificate of Completion has been issued, by the Engineer, for that part of the Works, including such a certificate for the whole of the Works. He is also responsible for the care of any work that he has undertaken to finish within the Maintenance Period, until such work is complete.

The Contractor undertakes to make good any damage to any work for which he has the responsibility for care, except where the damage has been caused by the 'Excepted Risks'. These are

> 'riot war invasion act of foreign enemies hostilities (whether war be declared or not) civil war rebellion revolution insurrection or military or usurped power ionising radiations or contamination by radioactivity from any nuclear fuel radioactive toxic explosive or other hazardous properties of any explosive nuclear assembly or nuclear component thereof pressure waves caused by aircraft or other aerial devices travelling at sonic or supersonic speeds or a cause due to occupation by the Employer his agents or servants or other contractors (not being employed by the Contractor) of any part of the Permanent Works or to fault defect or error or omission in the design of the Works (other than a design provided by the Contractor pursuant to his obligations under the Contract) (clause 20(3)).

clause 20(3)

Since the Contractor has the prime responsibility for the repair of any damage, unless this is caused by an exception, the cost will fall on him, initially. It will stay with him, unless he can prove that the cause of damage was one of the exceptions. The position is reversed 14 days after the issue of the Certificate of Completion, when the Employer has the prime responsibility for any damage to the work covered by the Certificate. Any outstanding work or maintenance being undertaken by the Contractor remains the responsibility of the Contractor (clauses 20(1) and 20(2)).

clause 20(1) clause 20(2)

It is obviously to the Contractor's benefit to obtain completion certificates for as much work as possible, particularly where others are in a position to cause damage. For example parts of carriageways on dualling projects or structures where processing plant is being installed (completion certificates are covered in chapter 3).

Of the excepted risks, the one most likely to be encountered is damage due to use or occupation, but there is always the possibility of a design fault or even acts of terrorism, which possibly would fall under acts of foreign enemies or rebellion or insurrection, depending on the particular circumstances. In Northern Ireland the Government has accepted the cost of repair due to terrorists, but the position is different in Britain.

The Contractor has the responsibility of repairing any damage caused by Excepted Risks but at the expense of the Employer. How this expense is to be calculated is not stated, but in the absence of any reference to either cost or Bill rates it is reasonable to assume that neither of these methods were intended, without the agreement of the parties. The charge to the Employer should be reasonable in all the circumstances, and should include overheads and profit. The consequential effect, if any, on the completion of other, undamaged parts of the Works should also be included.

Insurance of the Works

clause 21
clause 20

Clause 21 requires the Contractor to insure against all loss or damage, for which he is responsible, under clause 20, over the period of such liability. An exception to this requirement is made with regard to damage due to the incorporation of materials and workmanship not in accordance with the Contract, unless the Bill of Quantities contains a special item for its provision. Such an item is extremely unlikely, because the insurance is virtually unavailable, in the UK.

clause 21

The insurance of the Works is usually referred to as Contract All Risks, which is a bit of a misnomer, because, as has been noted, it does not cover all risks. Clause 21(b) requires the insurance policy to cover for the damage to Constructional Plant, but this may well be covered by a separate policy. The insurance policy, as referred to in clause 21, is an individual policy for the Contract, in the joint names of the Employer and Contractor. Many Contractors have a blanket policy, for all contracts, which is usually acceptable to the Employer, provided that his interest in the policy is acknowledged by the insurer.

Insurers need separate notification of some work situations such as working over water or between tides. The Site agent should be aware of these requirements and see that any additional work, involving the situations, are notified.

The amount the Contractor pays for insurance is likely to be based on his claims experience, the cover provided and the excesses incorporated in the terms. Premiums are not formally subject to a no claims bonus or surcharge, as is car insurance, but nevertheless the connection still exists. The amount of the premium for a policy for a single contract will be adjusted on the final value of the Works, and for extensions of time awarded, or prolongation on the part of the Contractor. The blanket policy premium will depend on the turnover, covered by the policy. The cover of the insurance policy will vary between insurers and contractors, so site will need a copy of this, or a schedule of the cover, prepared by the company's insurance department or brokers. The brokers advise on the company's insurance, and act as an intermediary between company and insurers when a claim arises, and may well provide a form for making a claim.

The one thing that is essential to receive, from site, because it is a term of all policies, is very prompt notification of any potential claims. This notice must not be put off because of a possibilty that the final claim will lie elsewhere. For example subsidence of a foundation could be caused by a design error, by the Engineer, or by poor compaction in an in situ pile. If it is a design error, there will be no insurance claim, but if poor compaction is the cause of the damage, an insurance claim should be submitted. If the damage was due to poor compaction in a pile, the repair of the faulty workmanship would not be recoverable from the insurers. In cases of doubt as to the cause, both the insurance brokers and the Engineer–Employer side should be notified of the possible claim, but the insurers should also be informed that the cause of the fault had not been established. Naturally the claim would only be paid once, when liability had been established. Prompt notification is even more important where the Contractor changes his insurers fairly frequently, looking for a better deal.

The value of any insurance claim will have to be agreed with the insurer's loss adjuster, which involves the agreement of a method of computing the cost, once the allowable heads of cost have been accepted. In the past loss adjusters have agreed to use the Federation of Civil Engineering Contractors' Daywork Schedule[4] as a basis, with a reduction to remove the profit element. The allowable heads of claim will depend on individual policies, particularly with regard to the consequential costs to the Contract work. The broker should be able to give guidance.

Damage or injury to persons and property and Insurance

Damage or injury

clause 22(1)

Clause 22(1) requires the Contractor to

> '... indemnify and keep indemnified the Employer against all losses and claims for injury or damage to any person or property whatsoever (other than the works) ... which may arise out of or in consequence of the construction of the Works and against all claims demands proceedings damages costs charges and expenses whatsoever in respect thereof or in relation thereto ...'.

clause 20

clause 22(1a)
clause 22(1b)

Other property of the Employer, where the Works are being executed in an existing complex, fall under this clause, not clause 20. There are a number of exceptions to the Contractor's obligation to indemnify the Employer, which are set out in clause 22(1) (a) and (1) (b). As the Contractor has the prime obligation, he would need proof that any injury or damage which occurred was caused by one of the exceptions, if he is to escape from the indemnity.

The Contractor should arrange with the Employer that all notices of injury or damage, received by the Employer/Engineer and not accepted by him as falling under one of the exceptions, should be passed immediately to the Contractor. This is not always done, because of the Employer's situation in the area in which the work takes place, and perhaps the fact that the clause does not require this to be done.

It is well worthwhile for the Site agent to aim at minimising claims, as far as possible, to try to maintain good relations with adjacent property owners and tenants, usually so necessary for the successful completion of the Contract. Many of the instances that are niggling and irritating to local people can be avoided. Examples include: lack of fencing, which allows plant drivers to take a more scenic or drier route; lack of sufficient, well serviced toilets, leading to the fouling of adjacent land; interference with services to adjacent properties; dust from earthworks operations which reduces the value of local crops; materials being left so close to boundaries that they can be sampled by inquisitive animals; gates on accesses being left open. Polythene is attractive to cows, particularly the very best milk yielders, and is lethal if eaten, since it blocks the stomach and intestines. New roads, pipelines, cables and drainage are the types of project which have a very long boundary with adjacent landowners and are also likely to have some accommodation works included in the overall Contract, and are thus potentially most liable to claims.

clause 22(1)

The exceptions listed in clause 22(1) (a) and (b), for which the Employer indemnifies the Contractor against any claims, should be noted, as they occasionally come into play.

clause 22(1) (b)

With respect to the events in clause 22(1) (b) (i)–(iv) the indemnity is unconditional but that under clause 22(1) (b) (v) is reduced proportionately to the extent that any act or omission of the Contractor contributes to any injury or damage. However, the Contractor should also note that the word 'unavoidable' is used in sub-clauses (i) and (iv) and the Contractor could have some liability, if only part of the damage was unavoidable. For instance some damage may be unavoidable owing to subsidence from tunnelling, driving headings, pumping or driving piles, as required by the Contract, but arguments are likely to involve the Contractor's precautions to limit such damage.

Public Liability Insurance

clause 22
clause 23

The Contractor is required to insure against any loss injury or damage to persons or property, but without limiting his obligations under clause 22 (clause 23). Such insurance is to run throughout the execution of the Works, which would include the maintenance. The insurance is referred to as Public Liability or Third Party insurance, and is required to be with an insurer, and in terms, approved by the Employer. The terms must include the insurer's giving the Employer the same indemnity that the Contractor is required to give, under clause 21. It is normal for insurance to be under the Contractor's blanket third party policy, subject to this covering the minimum amount stated in the Form of Tender (Appendix).

clause 21

Form of Tender (Appendix)

The premium paid is based on percentages of labour and supervision costs. The percentage varies with the amount of claims experience and the excesses, paid by the Contractor.

clause 22

The Contractor's liability under clause 22 is unlimited, even when there is a limit to the insurance cover he is required to take up under the Contract. It is for the Contractor to decide whether or not the contract minimum insurance cover is sufficient to safeguard his interests. As with insurance for the Works, the Contractor is required to produce the policy and a receipt for the current premium, whenever requested to do so.

Liability for the Contractor's employees

clause 24

Clause 24 refutes any Employer's responsibility for any damages or compensation, payable at law, in respect of any accident or injury to the Contractor's or his Sub-contractor's workmen. The Contractor is required to indemnify the Employer against such claims, except where the Employer, his agents or servants contribute to the accident or injury.

clause 21
clause 23
clause 22
clause 23

There is no clause similar to 21 or 23 that requires the Contractor to insure against claims from workmen, but he is required by law to do so. Perhaps unintentionally, the indemnity given by the Contractor under clause 22 includes claims for injury to workmen, but the insurance necessary to comply with clause 23 would not give cover for workmen's injuries. These would be claims on the employer's liability insurance. The premium is based on percentages of labour and supervision costs.

Contractor's design insurance

There is no provision in the ICE *Conditions* for the contractor to take out design (professional indemnity) insurance, but protection will be required if there is a considerable amount of temporary or permanent works design included in the Contract.

Sub-contractor's insurances

Sub-contractor's liabilities to the Contractor should be the same as those of the Contractor to the Employer, and the same types of insurance policy should be maintained, by the Sub-contractor. Checks that these remain in force should be made.

5. Valuation and payment

The clauses to which reference is made in this chapter are: 1, 5, 7(3), 10, 12(1) and (3), 13, 14(6), 17, 18, 20(2), 25, 26(1), 27(6), 31(2), 32, 36, 37(2), 38, 39, 40, 42(1), 44, 47(2), 49, 50, 51, 52, 53(8), 54, 55, 56, 57, 58, 59, 60, 63, 65, 69, 70, the Fluctuation Formula clause, Form of Tender.

General

The ICE *Conditions* are primarily designed for use where the value of the Works is to be computed by measuring the whole of the work actually executed, following the rules laid down in the Conditions. In the Form of Tender the Contractor makes the following offer

> '... we offer to construct and complete the whole of the said Works and maintain the Permanent Works in conformity with the said Drawings, Conditions of Contract, Specification and Bill of Quantities for such sum as may be ascertained in accordance with the said Conditions of Contract.'

clause 1(1)(i)

The sum mentioned in the Contractor's offer is the 'Contract Price' which, in clause 1(1)(i) is defined as

> '... the sum to be ascertained and paid in accordance with the provisions hereinafter contained for the construction completion and maintenance of the Works in accordance with the Contract'.

clause 1(1)(h)

clause 10

clause 47(2)

There is no Tender Total mentioned in the Form of Tender but this is defined in clause 1(1)(h) as 'Tender Total' means the total of the Priced Bill of Quantities at the date of acceptance of the Contractor's Tender for the Works'. After acceptance the Tender Total has no significance except that it is used to determine the amount of the Bond, should this be required (clause 10 and chapter 4) and in determining the reduced value of the Liquidated Damages should the Engineer issue any Partial Completion Certificates (clause 47(2)(b)(ii) and chapter 3).

Before acceptance, the Employer/Engineer may use the total of the Bill of Quantities to compare tenders, but this does not mean, necessarily, that a comparison of Contract Prices would bear a similar relationship, if the final quantities were to be different from those in the Tender Bill of Quantities. Although all tender sums received could be within a few per cent of each other, the range of individual prices in the various Bills of Quantities could vary by as much as 100%, or more. These differences come about from a number of causes including differences between contractors in methods of allocating costs, mistakes and weighting of items by some contractors to obtain a better cash flow or an enhanced final payment.

In order to get a better idea of the relative merits of the tenders to the Employer, the Engineer would have checked the effect of possible changes in quantities, or prolongation of parts of the Works, and possibly will have asked the Contractor to make certain changes before acceptance. The ICE's booklet *Guidance on the Preparation, Submission and Consideration of Tenders for Civil Engineering Contracts*[2] advises that such changes to rates should occur only in exceptional circumstances.

The Contract Price consists of the summation of a large number of elements which can be allocated to four broad categories

(*a*) those items which are measured and based on rates, whether directly from the Bill of Quantities or derived from those therein or fair rates; most of the work done comes within this category

(*b*) amounts based on actual cost incurred; they mainly involve work done or materials supplied by Nominated Sub-contractors and events disrupting the Contractor's progress

(*c*) daywork, which is partly based on actual cost and partly on notional cost, and is the method of payment for those variations which the Engineer considers cannot be measured and rated

(*d*) those operations carried out '. . . at the expense of the Employer . . .' consisting of dealing with antiquities (clause 32) and repairs of damage to the Works caused by the 'Excepted Risks' (clause 20(2)).

clause 32

clause 20(2)

Valuation based on rates

clause 56(1)

In accordance with clause 56(1)

'The Engineer shall except as otherwise stated ascertain and determine by admeasurement the value in accordance with the Contract of the work done in accordance with the Contract.'

The ascertainment of the value by admeasurement involves

- determination of the proper itemisation
- measurement to obtain the quantity of those items
- allocation of a rate for each item
- extension of quantity times rate to give the value of each item.

Itemisation

The Bill of Quantities is divided into sections and items in line with the method of measurement used in its preparation. In accordance with clause 57

clause 57

> 'Except where any statement or general or detailed description of the work in the Bill of Quantities expressly shows to the contrary Bills of Quantities shall be deemed to have been prepared and measurements shall be made according to the procedure set forth in the 'Civil Engineering Standard Method of Measurement' approved by the Institution of Civil Engineers and the Federation of Civil Engineering Contractors in 1976 or such later or amended edition thereof as may be stated in the Appendix to the Form of Tender to have been adopted in its preparation notwithstanding any general or local custom.'

The 1986 reprint of the ICE *Conditions* retains reference to the 1976 Method of Measurement although a revised method was published in 1985.[3] However, it is only to be expected that, for some time after the publication of a new method, Bills of Quantities prepared on the old method will still be included in new contracts.

It is necessary to check with the Appendix to the Form of Tender and the Bill of Quantities to ascertain which method was used for any particular bill. One should not assume that, where there is no entry in the Appendix to the Form of Tender, the 1976 Method of Measurement has been used, without first checking on the preamble to the Bill of Quantities.

Form of Tender (Appendix)

Form of Tender (Appendix)

Contracts with the Department of Transport and road or bridge contracts with local authorities usually incorporate Bills of Quantities based on the Department of Transport's Method of Measurement for Roads and Bridges.[5] Where this is so, it is normal for clause 57 to be revised accordingly. However, the method could be incorporated by a statement in the preamble to the Bills of Quantities.

clause 57

The incorporation of an entire method of measurement into the Contract should be fairly straightforward, but the billing of one or two items under a different method from the remainder of the bills often causes confusion and argument. If the change in method is deliberate there seems every reason for expressly stating so, but there are occasions when a mistake has been made, and is being defended. It is easy to understand mistakes being made when there are different ways of billing an operation in the various standard methods of measurement, and a person is taking off for the occasional civil engineering bill, when he normally works on building projects.

In the past there have been people caught out by the differences in the methods of measurement for a foundation excavation. The building method measured in layers, the civil engineering method from the surface to a range of depths and the roads and bridges method used just one item for an excavation. These methods are illustrated in Fig. 5.

The Contractor is entitled to request that any errors in itemisation in the Bills of Quantities should be corrected and the measurement of the Works made under the corrected items in accordance with clause 55(2) which states

clause 55(2)

> 'Any error in description in the Bill of Quantities or omission therefrom shall not vitiate the Contract nor release the Contractor from the execution of the whole or any part of the Works according to the Drawings and Specification or from his obligations or liabilities under the Contract. Any such error shall be corrected by the Engineer and the value of the work carried out shall be ascertained in accordance with clause 52. Providing that there shall be no rectification of any errors omissions or wrong estimates in the descriptions rates and prices inserted by the Contractor in the Bill of Quantities.'

This right to correction would not apply to any items, such as method related preliminaries, which the Contractor is entitled to insert in the Bills of Quantities.

The items contained in the bills may vary as a result of measured quantities differing from those billed (clause 56) and because of variations, instructed under clause 51, but these clauses are considered more fully under the section on rating.

clause 56
clause 51

Measurement

The quantities in the Bill of Quantities are only estimates and are not to be taken as the actual or correct quantities required to complete the Works (clause 55(1)). The actual quantities, on which the valuation is based, are ascertained and determined by admeasurement, by the Engineer or, if authorised by the Engineer, the Engineer's Representative (clause 56(1)). The Contractor is entitled to attend when the Engineer requires

clause 55(1)

clause 56(1)

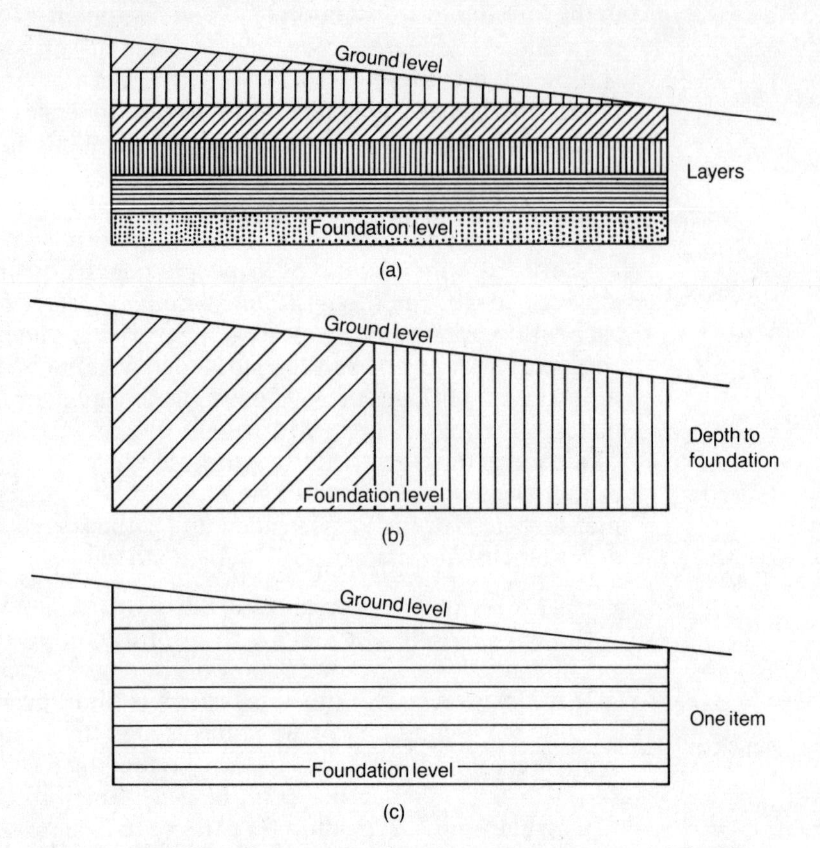

Fig. 5. Methods of measuring foundation excavation: (a) building; (b) Civil Engineering; (c) former Department of Transport method

clause 56(3)

any part of the Works to be measured (clause 56(3)).

It is important for the Contractor's measurement representative to be in attendance when work is to be measured, otherwise the measurements approved by the Engineer are taken as the correct measurements (clause 56(3)). Normally the Engineer's and Contractor's representatives will make mutually agreeable arrangements to measure the Works, where there is no particular urgency to do this. Where the opportunity to carry out site measurement is limited to the time between one operation and that following, the Contractor, by clause 38(1), is required to give the Engineer notice that the item of work is available for examination and measurement. The Engineer must then take his measurements without unreasonable delay. What constitutes unreasonable delay will depend on the particular circumstances. On an operation employing much expensive plant one would not expect to have to stop the plant working to allow the work to be measured, but, on the other hand, the Engineer could expect the work to be

clause 56(3)

clause 38(1)

organised to allow just enough time to allow measurements to be taken, section by section. The Contractor's notice that some section is or will be available for examination and measurement should also state when work on the section is to be continued, to make certain the Engineer is aware of the limits. Consideration is given in chapter 2 to the procedures for notices regarding the availability of work for inspection.

In addition to co-operation between the Contractor and Engineer, there is need for liaison between the Contractor's staff concerned with measurement and those responsible for the organisation of the work, on the larger projects where these may not be one and the same. This is particularly important where there is a non-standard method of measurement for an item, which requires site measurements when none would otherwise be required. For example, it is fairly standard for imported filling to be measured as the total volume of the embankment being filled, less the volume of suitable site excavation that is available. However, it is sometimes measured at the volume taken from the borrow pit. Even if the standard method of measuring imported filling is incorporated into the Bill of Quantities, the Contractor may also need to take measurements in the borrow pit to make payments to the owner of the pit. In such a case the Engineer would only be involved in the measurement in accordance with the Bill of Quantities.

The standard method of measurement gives the units of measurement for each item and also the limits of measurement or deductions from an overall measurement, for particular items, e.g. the length of a drain between manholes and the maximum size of openings that need not be deducted from a concrete slab, or from shuttering. The method also sets out the separate items necessary to bill adequately each section of the work, and the content of each such item.

If later time-consuming disputes are to be avoided, it is important for Contractor and Engineer to keep records in such a way that any later significant differences can easily be reconciled.

There are a number of ways that quantities can be obtained for a complete measurement of the Works, depending on the operation involved. The following are examples.

(*a*) Quantities necessitating Site measurements such as most earthworks quantities.
(*b*) Quantities which can be taken from drawings such as most concrete, shuttering and reinforcement for structures.
(*c*) Quantities which can only be measured from the

amount supplied such as rock filling into very soft ground, services of chainmen.

(*d*) Quantities, which are theoretical only, and are derived from calculations. The volume of imported filling based on the volume of embankment less the volume of excavated material, is an example. Because the measurement rule assumes that one cubic metre of excavation equals one cubic metre of filling, the amount of filling imported can be different to that measured for payment. Any difference can be accentuated if the Bill of Quantities is based on the Department of Transport's Method of Measurement for Roads and Bridges Works, because this does not require all available, suitable site filling to be added into the amount deducted from the total embankment volume.

Measurements in situ should be taken by the Contractor's and Engineer's representative together, wherever possible. In practice it is necessary to agree the volumes of items such as small soft spots which crop up irregularly during excavations between the foreman and the clerk of works who are always available at the site of the operation. In addition to the measurements, both the Engineer's and Contractor's record books of in situ measurements should include

- the names of the persons carrying out the measurement
- the date the measurements were taken
- the location of the items
- a complete description of the item—this should be what is seen on the ground, for a late comparison with the Bill of Quantities, thus providing a check on any differences, whether intentional or accidental, which may have been introduced
- a reference to any site instruction or variation order
- a signature from the person who has agreed the measurements.

In order to keep an overall record of what has been measured for such items as drainage, it is useful to mark the items measured on the drawings, giving the date measured and the location of the measurement record. This allows a visual check of those items which have not been measured. The basic requirement for all site records is that they can be read, understood and extended, if necessary, by another person in the absence of the originator.

Where site measurements are to be used in some formula or computer program to calculate the quantities, then it is of obvious advantage for Engineer and Contractor to agree, if possible, the method of calculation and the order of procedure, to ensure that documents are produced that are more easily comparable. Should mistakes creep into the compilation of total item quantity then reconciliation of figures is much easier, if comparison can be made at a fairly large number of stages. This is particularly true in the case of very large items such as earthworks. Where quantities are to be produced by taking them from drawings this will probably be done independently, by Contractor and Engineer. Nevertheless, some prior discussion and agreement as to layout and borders between items, such as that between concrete in walls and floors, can be an advantage to later reconciliation. If agreement on borders between items cannot be reached, the next best thing is to keep disputed quantities separately, to maintain easy comparison of overall quantity at a number of stages.

Both the Contractor and the Engineer will want to use measurements for both final and interim purposes. However, the Contractor should not be put off from continuing the process of measurement, when the Engineer or his representatives have other priorities. If it is difficult to agree on how to proceed, the Contractor should continue, and give a copy of his measurements to the Engineer to enable the Engineer to check, or at least follow, the same sequence. It is essential to follow this procedure as quickly as possible and preferably before the measured item is covered up, should any measurements have to be taken without the Engineer, in such circumstances.

Bill rates, adjusted rates and new rates

Having decided on the items and the quantities rates and prices have to be allocated to those items. If the Contract Documents have been well prepared, and few changes have occurred, most of the rates and prices will be those in the Bill of Quantities but in other cases there is a considerable potential for disagreement.

The items in a civil engineering bill are normally very brief and for the complete description the Contractor must look to the specification, drawings and the method of measurement. The method of measurement is a general document and the item coverage therein may be more comprehensive than that to be inferred from the item description in the Bills of Quantities, drawings and specification. For example, the Method of Measurement for Road and Bridge Works gives a square metre item for concrete paving to include all reinforcement, dowels, fillers, sealers and waterproof underlay, but, of these, only those inclusions detailed in the Contract Documents are priced in the Bills of Quantity rate. In such a case the tender rate applies only to the item coverage in the Bills, drawings and specification.

The bill rates apply to the descriptions available to the Contractor at the time of Tender and, in accordance with the Conditions of Contract, are subject to amendment only if affected by one or more of the following considerations.

clause 51
clause 52
clause 52(1)
clause 55(2)
clause 52

- (*a*) Work on variations issued under clause 51 and valued under clause 52. Consideration has to be given to whether or not the varied work is of similar character and executed under similar conditions to a bill item. If it is, then the billed rate or price is used. If it is not, then either the bill rates or prices are used as a basis for calculating a new rate, or a fair rate is applied (clause 52(1)).
- (*b*) Rates for new items introduced through the correction of errors or omissions in the Bill of Quantities. Consideration required as in (*a*) (clauses 55(2) and 52).
- (*c*) Rates for boreholes or exploratory excavation ordered by the Engineer, and either carried out as a variation or under an instruction against a Provisional Sum in the Bill of Quantities. Consideration as in (*a*) (clauses 18, 51, 52(1) and 58(7)).

clause 18
clause 51
clause 52(1)
clause 58(7)

- (*d*) Rates and prices for work, other than daywork, ordered to be carried out by the Contractor against a 'Provisional Sum' in the Bill of Quantities.
- (*e*) Rates or prices, which in the opinion of the Engineer have been rendered unreasonable by the 'nature or amount of any variation relative to the nature or amount of the whole of the contract work or to any part thereof' (clause 52(2)).

clause 52(2)

- (*f*) Rates, which in the opinion of the Engineer, have been rendered unreasonable or inapplicable because the actual quantities executed are greater or less than those shown in the Bill of Quantities (clause 56(2)).

clause 56(2)

- (*g*) Rates or prices for items which have not themselves been subject to a change in quantity, from that shown in the Bill of Quantities, but nevertheless have been affected by another item or items, which have been subject to such a change (clause 56(2)).

clause 56(2)

- (*h*) Rates and prices for items which are affected indirectly by an instruction to carry out boreholes or exploratory excavation as a variation (clauses 18, 51 and 52(2)).
- (*i*) Rates for work of rectification which, in the opinion of the Engineer, is not made necessary through the use of faulty materials or workmanship (clause 49(3)).

clause 18
clause 51
clause 52(2)
clause 49(3)

clause 52(1)

In arriving at rates or prices for the events given in items (*a*)–(*d*) and (*i*) above the Engineer is required to use the principles given in clause 52(1)—i.e. rates and prices to be those in the Bill, or based on those in the Bill, or if neither of these are appropriate, then they are to be fair rates and prices. There should be an attempt to agree rates by consultation between the Engineer and Contractor and, in the event of failure, the Engineer determines the rates and notifies the Contractor of his decision.

The rules for deriving rates for the situations in items (*e*)–(*h*) are even less clear than those for items (*a*)–(*d*) and (*i*). Rates rendered unreasonable by a change in quantities to those Billed, are to be adjusted by an appropriate increase or decrease, after consultation between Engineer and Contractor ((*f*) and (*g*)). Rates indirectly affected by variations and instructions for boreholes are to be fixed at a level which the Engineer considers to be 'reasonable and proper', with no consultation mentioned (items (*e*) and (*h*)).

Although the wording in the ICE *Conditions* with regard to the fixing of rates is not consistent there is a tendency to try to use rates derived from bill rates wherever possible. If this cannot be done there seems little difference between a fair rate and one that is 'reasonable and proper'. There is no definition in the ICE *Conditions* of either a 'fair rate' or one that is 'reasonable and proper'. A commonsense interpretation of the terms is the provision of rates which give the Contractor a reasonable profit, after taking into account oncosts and overheads. This involves basing the operation cost on the reasonable cost anticipated, if the rate is being fixed in advance of the work, or in other cases that incurred.

If the individual operation rates include allowances for oncosts and overheads and profits, it is normal to add the percentage for these, contained in the bill items, to the operation cost to obtain the new rate. If some or all of the oncosts and so on are included in preliminary items some adjustment may need to be made to these items. This adjustment is normally made to cover a number of changes rather than each individual one.

When the Engineer fixes a rate or price, the Contractor, if he is dissatisfied, has 28 days to notify the Engineer, in writing, that he intends to claim a higher rate. (Chapter 6 deals with the claims arising from such a notification.) It is important for the Contractor to be able to differentiate between communications that are a part of consultations to arrive at a rate and those fixing the rate. It has been known (although not often) for the Engineer to include rates in the variation instruction, and then argue that this was notification of the fixed rates. This unusually autocratic behaviour can sour relationships, but so can letters claiming this, that, and the other, when no rate has been fixed at that point. If a form of wording—to be used only when a letter fixes a rate—is agreed beforehand, there will be no excuses possible for missing the 28 days deadline, within which notice of a claim has to be given.

The agreement of rates for variations should be carried out at regular intervals, at meetings set up for this particular purpose. Meetings would be more productive if proposed rates, together with their make-up were exchanged before the meetings, but this is by no means always the case. Some Contractors and Engineers like to play their cards very close to their chests, maintaining as much as possible in play, until all the factors affecting a decision are known. If there is no similar rate in the Bills for a variation then as a first choice the bill rates are used as a basis for a valuation of the variation. The most general way of doing this is to establish the formula, used in the Tender, for the build up of the rates in the Bills, such as the following:

Rate = anticipated cost of the operation + particular and general site oncosts + off-site overheads + profit

The formula can be made more complicated by splitting the costs into labour, plant, materials and sub-contractors, each of which can have a separate addition for overheads and profit. It may be necessary also to have separate formulae for each section of the Bill. The oncosts, overheads and profit would normally be expressed as percentages, but this is only a true relationship in some cases and where other parameters do not change.

The higher the percentages put forward by the Contractor for the additions to cost, the more proof he will be required to produce to substantiate his figures. The proof most usually required is that the basic costs after the removal of the uplifts are adequate. The anticipated cost should be based on the cost of resources at the time of tender, if the fluctuation clause applies to the Contract, otherwise it will be an average cost over the relevant period for each item. The use of the formula for new rates, given above, would need to account for inflation in one way or another. In some instances the revised anticipated operation cost can be obtained by applying a pro-rata to the relevant part of the original anticipated cost, e.g. if a variation reduces output on an item by one third, the cost of labour and plant are increased by one half. In other cases the revised anticipated operation cost may have to be derived from the cost of resources to be applied and the anticipated output. Before any pro-rata can be applied a further breakdown of the elements of cost will probably be required, to isolate that part which is affected by the variation. One example is an increase in the cement content of concrete which requires the material cost to be broken down into separate costs for cement, sand aggregates and additives.

Determination of a new rate for an item for selective filling to an earth embankment, where compaction additional to that specified is instructed, would involve breaking down the labour and plant element into spreading, levelling and compacting, before applying a pro-rata adjustment. There is the possibility that the increased compaction will restrict the output of the filling to that which can be satisfactorily compacted, which will entail the adjustment of the remaining constituents of the labour and plant. The greater compaction could also affect the balance of earthworks quantities, and the ratio between the amount of imported fill used and that included in the measurement. There are a number of factors that can reasonably delay the agreement of a rate for a variation, such as the conglomerate items, which often occur in civil engineering bills. For instance, an earthworks item on a road contract may have an average haul of 600 m, but hauls may vary from a few metres to several kilometres. A variation that added or omitted material within that range of hauls may well be described as of similar character and executed under similar conditions as the work billed, but may affect the eventual average haul. The calculation of a new rate would not just involve a pro-rata based on the anticipated and actual average hauls, because in most cases the same plant would not be used over the full range of hauls. However, a number of variations, some of which are of material within the original range of haul distances, and some outside that range, may cancel each other out as far as an effect on the average haul is concerned. The more simple items, where the components of the average rate, inserted in the bill, do not vary too much from that average, are easier to deal with at the time they occur. In order to base a rate on bill rates where the varied work is not of similar character or carried out in similar conditions to the work billed, the original components of the bill rate must be revealed by the Contractor. If this is not done then the Engineer can only assume values for those components, and the Contractor has no reasonable complaint if the rate fixed is not to his liking.

Most examples of the problems which can arise seem unreal, but one true instance regarding the pricing of a large variation order for hardcore material caused a dispute which went on well after work was completed. Neither the Contractor nor the Engineer had supplied a detailed breakdown of the new rate to the other, and so the discrepancy between the two versions was not revealed. It was found out much later that because the material was being purchased by volume in the lorry, the Engineer's Representative had for pricing purposes taken the material laid as that in the lorry and had omitted the compaction factor. However, revealing a make-up of a rate which includes an error can involve more trouble than refusing to reveal the make-up, but Contractors should not allow such an experience to put them off.

clause 52(1)

The other method of arriving at a rate, where the bill rates are neither suitable for direct use nor as a base for deriving new rates, is to determine a fair rate. As clause 52(1) is written a fair rate can only be used if it is not reasonable to apply the other methods. It is not an alternative for them. If a fair rate and a rate based on bill rates were more or less one and the same thing then the precedence of one method of determination over another would have no effect. However, a fair rate is usually accepted as one that gives the Contractor his reasonable costs, including overheads, plus a reasonable profit, whereas a rate based on bill rates could give rise to a loss or an unreasonable profit, depending on the bill rate.

clause 52(2)
clause 56(2)

Clauses 52(2) and 56(2) deal with the indirect effect of variations and changes of quantities on other items in the Bill of Quantities. The Contractor is required to give notice to the Engineer, or vice versa, if either believe that any rates will be affected. The notice is to be given, in the case of variation

> '. . . before the varied work is commenced or as soon thereafter as is reasonable in all the circumstances . . .'.

clause 56(2)

clause 56(2)
clause 51(2)

In the case of increases of quantities being the cause of the indirect effect there is no time limit for the notice, given in clause 56(2). This may arise because the difference in quantities may not be discovered until all measurements have been established. There seems to be no reason why amended rates under clause 56(2) are determined after consultation with the Contractor, but consultation is not a requirement of clause 51(2). In practice there is usually no differentiation.

The following are examples of the indirect effect of variations and changes on other rates.

(*a*) A variation to finish shuttering increases the time required to fix and strip shuttering for each lift on a concrete chimney. The cycle time for shuttering, reinforcement and concreting will also be increased by the same amount, thus prolonging the overall time for construction, and reducing the average output per hour of concrete and reinforcement. The Contractor would be entitled to request adjustment of the rates for concrete and reinforcement, and any rates or prices containing time related overheads, although none of these items were mentioned in the variation order.

(*b*) An instruction to deepen the excavation for a structure, which is being carried out with battered sides, may involve the use of a larger crane to lift in materials used in construction, because of the greater reach involved. The size of pump required to keep the excavation dry could also be affected. Any items containing an allowance to cover the cost of the crane and pump would need to be adjusted.

(*c*) One or more variations that prolonged one part of the project, without the need for an overall extension of time, would affect all rates that contained an element for time related overheads for that particular part of the project.

These examples show that the main indirect effect to be taken into account, when adjusting the rates for one or more other items, is the delay and disruption caused to other work. The delay may or may not justify an extension of time under clause 44(2), but nevertheless its effect on other rates has to be considered and those rates affected adjusted in accordance with clause 52(2).

clause 44(2)

clause 52(2)

Clause 44 gives the Contractor an entitlement to an extension of time if variations or increased quantities delay the completion of the Works, but that clause gives the Contractor no entitlement to additional payment for the delay—the reason being that clause 44 covers all delays leading to extension entitlement, irrespective of the position regarding reimbursement to the Contractor for the consequences of that delay on his finances.

clause 44

clause 44

It is clause 52(2) and 56(2) which give the Contractor the right to reimbursement for the delays and disruption to part of the Works or the whole of the Works due to variations through the adjustment of any affected rates. The general situation with regard to reimbursement for delay is that this occurs only if the delay is deemed to be within the control of the Employer, his servants or agents. However, other circumstances giving rise to reimbursement can be written into the Contract. In the case of the ICE *Conditions* the costs of any delay arising from physical conditions or artificial obstructions not anticipated by

clause 52(2)
clause 56(2)

an experienced contractor can be recovered, although it is outside the control of the Employer (clause 12).

clause 12

The Civil Engineering Standard Method of Measurement (CESMM) allows the Contractor to insert method related items into the Bill of Quantities at the time of Tender, but does not make this mandatory (CESMM Class A—General Items and Section 7—Method Related Charges). As a consequence the way that the Contractor's general charges are included in a bill of quantities varies from almost complete inclusion in the work items to almost complete listing as 'Method Related Charges' in the General Bill. In addition there can be a mixture of the two methods of including these charges.

Where all or part of the Contractor's general charges are included as a proportion of the rates for work items, this may be as a single percentage over all items or the percentage may vary over the different sections of the bills. There can be a number of reasons for the varying percentage, including the following.

- An error in a section was discovered too late to allow correction of individual rates, and the error was compensated for by the addition or subtraction of an equal total amount, from an item in the general section of the bill.
- A similar amendment to that mentioned in the above example would be required should the Contractor receive a subcontract tender that he wanted to include in his tender, shortly before that tender was due to be returned.
- The Contractor might add a greater proportion of overheads to the items of work to be done early in the programme, in order to obtain a cash flow advantage and reduce the amount of his own money tied up in the project.
- The Contractor might add weight to the rates for some items where he thought the quantity was likely to increase or conversely reduce those items where the quantities were expected to decrease.

The distortion of the bill rates, for reasons similar to the ones quoted, makes the task of agreeing adjustments to rates affected by variations a much more difficult one. If significant adjustments need to be made, the Engineer may ask for a complete breakdown of a section or the whole of the Bill, in order to ensure that all the necessary allowances had been included, and were reasonable. If only partial breakdowns are given there may be the suspicion that these have been distorted to justify the increase requested.

Cost items

The second part of the valuation is that based on cost, or additional cost to the Contractor. This is mentioned in item (*b*) of the summary on p. 79. These items include those for which some specific provision is made in the Bill of Quantities, mainly nominated Sub-contractors, and those which can only be covered by a Provisional Sum for a general contingency allowance. These latter items can range from incidental costs arising during construction, to compensation arising for unanticipated events that cause delay or disruption to the progress of the Works. The entitlement to recover the cost of these items comes from provisions contained in the ICE *Conditions*.

Items for which provision is made in the Bill

The items for which provision is made in the Bill are to be carried out by Nominated Sub-contractors, which, in the ICE *Conditions*, is a term applied to firms or persons nominated by the Engineer to supply either goods or services or a combination of both. There are two ways of incorporating Nominated Sub-contractors in the Bill of Quantities, as follows.

clause 58 (2)–(6)

(*a*) Inclusion of a Prime Cost Item where the Engineer retains the right to instruct the Contractor to employ a sub-contractor, of the Engineer's choice, to supply goods or materials or execute part of the work: (clauses 58(2)–(6) and also clauses 5.15 and 5.16 of the *Civil Engineering Standard Method of Measurement*). Each Prime Cost item in the bill is replaced, in the valuation, by the actual cost to the Contractor of the Sub-contractor's supply, based on the particular sub-contract. In addition, the Contractor is entitled to the allowance for labours which he has inserted in the Bill, plus the percentage of the actual Sub-contractor's cost that the Contractor inserted against the relevant bill item, to cover for all other charges and profit. If the bill contained no item for percentage, for profit and so on, the percentage inserted by the Contractor in the Form of Tender (Appendix) would be utilised in the valuation.

Form of Tender (Appendix)

clause 58(7)

(*b*) Inclusion of a Provisional Sum for materials or work which will be required to be supplied or executed, only if instructed by the Engineer: The Engineer, by clause 58(7) is entitled to nominate a sub-contractor to carry out this supply.

In such cases the way of payment is the same as if

Form of Tender (Appendix)

there was a Prime Cost Item, but in the absence of an item for the Contractor's labours, a rate for these will be agreed or fixed by the Engineer, and the percentage in the Appendix to the Form of Tender will be utilized to cover 'all other charges and profit'.

clause 59A(5)

The cost of the supply included in the valuation should be net of all discounts, except any discount for prompt payment, which the Contractor has been able to agree with the Sub-contractor (clause 59A(5)(a)). The Sub-contractor is under no pressure from the ICE *Conditions* to give such a discount, in the respect that the Contractor is not entitled to object to the nomination, if no discount is offered.

Costs for which items in the Bill are unlikely

The ICE *Conditions* provide for the payment to the Contractor of amounts based on cost if certain situations or events occur during the course of the Contract. The recovery for these situations can come under three categories

(*a*) recovery of cost plus overheads plus profit (COP)
(*b*) recovery of cost plus overheads (CO)
(*c*) recovery of net cost only (C).

There are other ways of categorising these situations, such as the seriousness of the effect they might have on the overall project, but some remarks on this are made, and the more serious effects are considered, under Claims in chapter 6.

clause 1(5)

In accordance with clause 1(5) 'cost' is deemed to include '. . . overhead costs whether on or off the Site except where the contrary is expressly stated.' Therefore where the word cost is used without further description, in the ICE *Conditions* the situation falls under category (b).

The provisions for situations likely to effect the Contract seriously include reference to 'delay and cost' (DC). This 'cost' would include the cost of delay and disruption caused to any part of the Works, caused by the particular situation. In addition some clauses mention disruption separately.

clause 7(3)

- clause 7(3): delay in the issue of Drawings, which the Contractor has requested (pp. 24–25, chapter 2) (CO) (DC)

clause 12

- clause 12: occurrence of 'Adverse Physical Conditions and Artificial Obstructions', which an experienced Contractor could not have foreseen at the time of the Tender (COP) (DC)

clause 13(3)

clause 5

clause 13(3)

- clause 13(3): instructions of the Engineer with regard to explaining or adjusting ambiguities or discrepancies in the documents under clause 5, or any other instructions issued by him which complied with the requirements of clause 13(3) (CO) (DC)

clause 14(6)

- clause 14(6): unreasonable delay by the 'Engineer, in approving the Contractor's method of working, or the imposition of restrictions on that method due to design considerations of which, an experienced contractor could not have been aware, at the time of Tender (CO) (DC)

clause 17

- clause 17: rectification of any errors in setting-out, caused by incorrect data, supplied in writing by the Engineer. (CO)

  This clause gives the Engineer no authority to sanction payments for delay, caused by the rectification of errors in setting-out. Where this occurs reimbursement of the cost of that delay could be sanctioned, either by treating the instructions that contain the correct setting-out data as an instruction issued under clause 13(1), or a variation, issued under clause 51(1)

clause 13(1) clause 51(1)

clause 26(1)

- clause 26(1): the recovery of any fees, rates or taxes levied on the Site, or structures erected thereon, or temporary structures erected elsewhere, and properly paid by the Contractor (C); rates have not been charged on site offices which are only used for a few months but could be imposed for longer periods

clause 27(6)

- clause 27(6): cost of delays to the progress of the Works attributable to the need to give at least 21 days' notice before carrying out variation works 'in a Street or in Controlled land or in a Prospectively Maintainable Highway', (CO) (DC)

clause 31(2)

- clause 31(2): cost of any delays to progress caused by the Employer's workmen or his other contractors or properly authorized authorities or statutory bodies, beyond that which could have been reasonably foreseen by an experienced contractor at the time of the Tender (CO) (DC)

clause 36(2)

- clause 36(2): cost of any samples other than those intended by, or provided for in, the Contract to be at the cost of the Contractor (CO)

clause 36(3)

- clause 36(3): the cost of any test, other than those intended by, or provided for in, the Contract, to be at the cost of the Contractor, provided such test does not reveal workmanship or materials not in accordance with the Contract (CO)

clause 36(3)

- clause 36(3): the cost of any test under load or test to ascertain that the design of any part fulfils the intended purpose, provided that such a test does not reveal workmanship or materials not in accordance with the Contract (CO)

clause 38(2)

- clause 38(2): the cost of uncovering or opening up work and any necessary reinstatement, provided the Engineer's approval was obtained before covering up, and no workmanship or materials not in accordance with the Contract are revealed (CO)

clause 40(1)

- clause 40(1): the cost of any order from the Engineer to suspend any part of the Works, unless it is provided for in the Contract, is necessary because of weather conditions, or by some default of the Contractor, or is necessary for the proper execution or the safety of the Works providing the necessity does not arise through default of the Engineer/Employer or through one of the Excepted Risks (CO) (DC)

clause 42(1)

- clause 42(1): the cost of any delay as a consequence of the Contractor's not being given possession of any part of the Site in accordance with his programme (CO) (DC)

clause 50

- clause 50: the cost of searching for any defect imperfection or fault in the Works unless the problem is the responsisbility of the Contractor under the Contract (CO)

clause 59B(4)

- clause 59B(4): cost of any delay caused by the proper giving of notice of termination to a Nominated Sub-contractor. Proper notice is one given with the Employer's/ Engineer's consent or which the Contractor is entitled to give in accordance with the Sub-contract (CO) (DC)

clause 69

- clause 69: labour tax matters (C)

clause 69

If the Contract Price fluctuation clause is included in the Contract, clause 69 only applies to the Contract in so far as any matter dealt with therein is not covered by the *Index of labour in civil engineering construction.*

The labour taxes, for which the Contractor makes allowance in his tender rates and prices are those which he was actually liable to pay on the date for return of tenders, which is referred to in the clause as 'the relevant date'. The taxes coming within clause 69 are

clause 69

> '... taxes levies and contributions (including national insurance contributions but excluding income tax and any levy payable under the Industrial Act 1964) which are payable by the Contractor in respect of his workpeople and the premiums and refunds (if any) which are by law payable to the contractor in respect of his workpeople.'

At the time of writing only national insurance contributions seem to come within the ambit of this clause, and the clause, as written, does not fit well the present method of calculating the amount of this tax. The valuation is adjusted if there is any change in level or incidence of any labour tax matter from that which was levied, from employers, on the relevant date (clause 69(2) and (3)).

clause 69(2)
clause 69(3)

The clause was written when the employer's contribution for national insurance, for each person employed, was a defined amount per week. At that time 'level', in the last excerpt of the clause, referred to this amount. In the past an increase in wages tended to be followed by an increase in the level of the national insurance contributions. As at mid 1989, the tax is a percentage of wages and level must refer to this percentage. The amount paid will increase when wages increase, but the level will remain unchanged. Changes in the upper and lower limit of wages, on which the tax is charged, would represent changes in the incidence of the tax. At the time of writing the upper limit has been abolished, and only the lower limit remains.

The *Index of labour in civil engineering construction* does include allowance for the employer's national insurance contributions. However, because the amount of the wages allowance in the index—on which the national insurance percentage is charged—is between the upper and lower limits, there is no cover for changes of incidence of the tax (see *The price adjustment formula for civil engineering contracts*, obtainable from HM Stationery Office) but the index does reflect any additional amounts of insurance contributions when the wages are increased. The amount of adjustment to the valuation is the net amount of any increases or decreases consequent on a tax change, but only the contributions in relation to persons employed on manual labour on the Site are included in this adjustment (clause 69(4)).

clause 69(4)

clause 69(5)

Clause 69(5) sanctions the inclusion in Nominated Subcontracts of a similar provision allowing recovery of tax changes, but the wording of the clause does not allow the valuation to be adjusted in respect of tax changes, affecting other sub-contractors.

clause 52(4)

Notification and detailed description of such events that lead to an adjustment of the valuation is required, as they occur in accordance with clause 52(4). Such reports are designated as 'claims', although there is little reason for many of them to become contentious. Claims, likely to prove contentious are dealt with at greater length in Chapter 6.

Dayworks

clause 52(3)

Under clause 52(3)

> 'The Engineer may if in his opinion it is necessary or desirable order in writing that any additional or substituted work shall be executed on a daywork basis.'

Allowance for some daywork is made by means of the inclusion of a provisional Sum in the Bill of Quantities. The method of valuation for daywork is based on the hours of labour and plant properly used on the work, but the rates applied to those hours can be nominal, such as those from a schedule of rates, which gives rise to a notional rather than an actual cost. Materials are based on the actual cost plus a percentage for overheads and profit.

clause 52(3)

The provisions for daywork in clause 52(3) offer alternatives for the person preparing a daywork section in the Bill of Quantities.

(*a*) The simpler of the alternatives is just to provide a Provisional Sum for Dayworks, with no further details, in which case the hours and materials supplied will be priced out and included in the valuation in accordance with the rules and prices set out in the 'Schedules of Dayworks carried out incidental to Contract Work' issued by The Federation of Civil Engineering Contractors current at the date of the execution of the work'. (FCEC schedules).

If the Bill of Quantities makes no reference to Dayworks, any ordered should be valued using the FCEC schedules. Since, in this case, there would be no Provisional Sum, and therefore no budgeted allowance in the Bill of Quantities for Daywork, the Engineer might need to give the Employer some explanation.

(*b*) The other method requires the inclusion of a Daywork Schedule giving classes of labour and items of plant in the Bill of Quantities, with a provisional number of hours against each of these, to which the Contractor puts his own rates and prices. Thus some competition is introduced into the daywork section.

These schedules are seldom comprehensive but may provide a basis for deriving any further rates needed during construction.

Competitive pricing by Tenderers of a daywork section of the Bill of Quantities can be obtained by retaining the FCEC Schedule as a basis for payment but with provisions for each Tenderer to state his individual adjustment percentages on the amounts due for labour, plant, materials and Supplementary Charges.

The value of the daywork is then priced out in accordance with the FCEC Daywork Schedule and the totals of labour, plant, materials and supplementary charges increased or decreased by the respective adjustment percentages. This is set out in the *Civil Engineering Standard Method of Measurement*[3] (CESMM), as an alternative to the use of an itemized schedule (paragraph 5.6 of section 5 and item 4 under class A—General items). The Standard Method of Measurement does not contemplate the use of the FCEC Dayworks Schedule, without the opportunity for adjustment.

Sometimes arguments arise as to whether the percentages against the provisional sums are meant to replace the percentages contained in the FCEC Daywork Schedule, or are additional to these. The CESMM, in section 5.6(b), clarifies this situation by calling the percentages in the Bill of Quantities 'further adjustments'. However, the answer could be different if the Contract Bill of Quantities was not based on the CESMM.

In using the CESMM alternative method for inclusion of dayworks' charges it should be noted that certain of the Supplementary Charges (i.e. those under items 2(ii), 4 and 7 of Section 4 of the FCEC Schedule), are not included under that heading in the valuation, but under Labour, Plant or Materials, as appropriate, and as noted against those items. The note in section 5.6(b) of the CESMM does not mean that those items are to be covered by the percentages.

Keeping Daywork records

clause 52(3)

Clause 52(3) requires the Contractor to provide the Engineer's Representative with a daily record of the resources used on work which the Engineer has instructed to be carried out on a Daywork basis. The records should be

> '. . . an exact list in duplicate of the names occupation and time of all workmen employed on such work and a statement also in duplicate showing the quantity of all materials and plant used thereon or therefor (other than the plant which is included in the percentage addition in accordance with the Schedule under which payment is being made). One copy of each list and statement will if correct or when agreed be signed by the Engineer's Representative and returned to the Contractor. At the end of each month the Contractor shall deliver to the Engineer's Representative a priced statement of the labour material and plant (except as aforesaid) used and the Contractor shall not be entitled to any payment unless such lists and statements have been fully and punctually rendered . . . '

clause 52(3)

The Engineer has discretion to allow payments for Daywork, if late records are received, but only if he considers that it was impracticable for the Contractor to comply with the clause 52(3) requirements, and then only to the extent that he can satisfy himself as to the correct amount. A prudent Contractor should assume that the Engineer will insist on absolute compliance with the procedure in clause 52(3).

Pricing Dayworks Records on the FCEC Schedule

The Dayworks should be priced on the Schedule current at the time the Dayworks were executed, so it is important to ensure that the correct schedule is being used. The FCEC issue a number of amendment sheets to the schedule, in between complete reprints, and it is these that are most likely to be missed. It is not necessarily sufficient to have the latest amendment, as an earlier issued change may also apply to the printed edition. The Daywork Schedule contains four sections

- Labour
- Materials
- Plant
- Supplementary charges

Each section contains explanatory notes to help in the pricing of daywork charges, which it is essential should be fully understood.

The notes on labour give the content of the 'Amount of Wages' and the percentage addition to these. A problem that arises is in connection with incentive bonus paid to operators, working on daywork, which despite the reference in note 1 of the FCEC Schedule to 'actual bonus paid' is queried. The basis of the query is that the only definite bonus required to be paid in accordance with the Working Rule Agreement is the 'Guaranteed Minimum Bonus' although incentive bonuses are allowable under the agreement. The intention of the wording is as stated—i.e. bonus paid should be included in the amount of wages.

It can be agreed that overtime rates are payable based on the time of day that the daywork was executed, in which case the time of day will be required to be stated on the records, where less than a full day is worked. As an alternative agreement, the hours worked on daywork can be enhanced by multiplying them by the total paid hours divided by the total number of hours worked for the week in which the daywork was executed. The first method is easier. If the rates charged for a particular operative do not fit the description of that employee's trade, then queries can arise at a later date from auditors. A little care can avoid the necessity to dig out records after the end of construction.

Note 3 in section 2 regarding the taking of materials from Site stockpiles is often missed, in pricing the records.

In respect of plant the minimum charge is for the period stated in the schedule and the cost of fuel distribution is an additional charge (notes 2 and 3 in section 3). It is handy to ask the plant department to supply the rating of the items of plant in the fleet, on Site. If plant is rated by volumetric capacity the rating is based on the maker's capacity which may not be the size of bucket fitted at the time. Category 42 of the plant section does contain a number of items which can be omitted from the record sheets.

Particular consideration should be given to any supplementary charges that may apply, because items such as welfare, transport to site, and so on will probably not appear on the records (section 4 of the schedule).

Work at the expense of the employer

The wording 'at the expense of the Employer', in respect of the amount of the Contractor's entitlement to payment, appears only twice in the ICE *Conditions*

clause 20(2)

(*a*) under clause 20(2) the repair of any damage to the works, caused by one of the 'excepted risks' which has been instructed by the Engineer, is 'at the expense of the Employer'

clause 32

(*b*) under clause 32 the disposal of any fossils, antiquities and so on, on the instructions of the Engineer, are carried out 'at the expense of the Employer'.

clause 52

There is no reference to either of these items of work being variation orders, which would have brought the valuation of them within clause 52, and no clarification of the phrase 'at the expense of the Employer'. Presumably then, the Employer will pay a reasonable market price for the work, which will not necessarily be related to the Bill rates and prices.

clause 32

clause 40

With regard to the fossils, antiquities and so on it is only the disposal which comes within the ambit of clause 32. Other instructions connected to the discovery of the articles would be valued under some other clause, e.g. if work had to be stopped while some archeological remains were examined in situ, a suspension order, under clause 40, would be necessary.

Interim valuations certificates and payments

Form of Tender (Appendix)

The Contractor is entitled to interim payments at monthly intervals, for work done, materials supplied and other entitlements, provided that the amount due exceeds the minimum interim certificate amount stated in the Form of Tender (Appendix) (clause 60(1) and (2)). Clause 60 refers to work done up to the end of each month, but this does not necessarily clause 60(1) clause 60(2)

mean the end of a calendar month. A contractor usually prefers to have a month's ending coincide with the end of a week, which simplifies the comparison of costs and income.

The list of month end dates should be agreed at the beginning of the Contract, and would normally be the nearest week-end to the calendar month end. Exceptions might occur at holidays, end of financial years and, in the case of some local authorities, to ensure that the certificates would be available by the relevant committee meeting dates if interim payments had to be approved before payment could be made. The Contractor should try to foresee any such problems that affect payment, as if he does not his cash flow will suffer through late payment. The fact that the Employer is responsible for the delay in payment, and will be obliged to pay interest under clause 60(6), may not be sufficient to compensate for the lack of cash.

clause 60(6)

It is the Contractor who initiates the payment by submitting

'. . . a statement (in such form if any as may be prescribed in the Specification). . . '.

clause 60(1)

The contents of this statement, given in clause 60(1), are the cumulative estimates of all the elements of the valuation, detailed earlier in this chapter, plus the value of allowable materials, up to the end of the month. More often than not, the form of the statement is not mentioned in the Appendix, but a fully detailed valuation to date—showing quantities against each rate and full details of any other amounts—is a normal requirement. Clause 52(4) (d) and (f) requires provision of sufficient interim details of any claims, before any amount for these can be included in a certificate. The amount of detail required might not be the Contractor's idea of an estimate, but to insist on submitting less detailed statements may sour relationships and result in delayed payments. Nevertheless, the Contractor should protest against some of the more outrageous demands, such as the submission of nine copies of a detailed, typed, interim statement, unless this is a requirement of the Specification.

clause 52(4)

If the Contractor has not managed to produce full details of claims, up to submission of an interim statement, it is useful to include amounts in the statement as an indication of the anticipated value. This, if nothing else, does allow the Engineer/Employer to budget for future payments, which is a most important precedent condition to allow that payment. At the same time the Contractor's head office should be advised that these amounts are not expected to be paid until the details have been produced.

The timing for the submission of this statement, calculated to avoid problems of late payment, should be strictly maintained, as the next two steps in the payment procedure have to be completed within 28 days of the delivery, to the Engineer, of this application (clause 60(2)). The first of these steps is for the Engineer to certify the amount to be paid, and the second is for the Employer to make payment.

clause 60(2)

Preliminary agreement as to how certain elements of the tender are to be included in interim certificates will help to avoid delayed payments. If the Engineer disagrees with the method put forward by the Contractor in an interim statement, he may decide that there is no time for discussions, and would certify less than he would do after consultation.

Preliminary and temporary works items inserted by the Contractor in the Bill of Quantities as lump sums can be described as method related items, in accordance with the CESMM, page 13, which should simplify any agreement of a method of including them in interim certificates. They will either be one-off payments, or related to time. Those items related to time are not necessarily spread out over the full construction period, but may only be related to one section or even one operation. The Contractor should indicate the relevant period, for each time-related preliminary, but to do this thoroughly requires a large number of items, so it is not often done.

Where sums are entered in the preliminaries not as method-related items the Contractor should beware of putting forward some method of claiming payment for these on an interim basis that reflects the wrong relationship with the way cost is incurred. This may be put forward, by the Engineer, to the Contractor's disadvantage at some later stage when the correct relationship is included in a claim.

The CESMM allows the use of an adjustment item, which is used to make allowance for any late changes to the Tender. The interim payment for this item is always based on value, but total payment can neither exceed nor be less than the total of the bill item (page 12 of the CESMM).

The materials, for which payment is made on an interim basis, are materials for the Permanent Works, delivered to and stored on the Site, and those listed in the Appendix to the Form of Tender, which are stored off Site. A problem can arise when materials are stored on an area that the Contractor has arranged to rent or buy, from a third party, as an office site. Even though such an area is adjacent to the Site, technically, it will

Form of Tender (Appendix)

clause 1(n)

not be part of it (see the definition of the Site in clause 1(n) of the ICE *Conditions*).

Form of Tender (Appendix)

To have materials stored off Site included in an interim certificate requires the Employer's consent, unless the particular materials are listed in the Form of Tender (Appendix). It should be easier to come to some agreement, before any items are offloaded, unless there is absolutely no available storage area on Site. After all, Employers can be as aware of the advantages of cash flow as are Contractors. Any agreement would need to give the Employer the same protection as that for other off-site materials, provided for in the Contract.

clause 54
clause 60(1)
clause 54

Before any payment can be made for any materials stored off site, the property in those goods must be vested in the Employer, in accordance with clause 54 (clause 60(1) (c)). The action that the Contractor is required to take is set out clearly in clause 54, but the Contractor should realise the problems that may arise at a later date.

In most cases of payment for materials stored off site, the materials are on the premises of a supplier, often a nominated sub-contractor, who has been paid by the Contractor. The inspection of materials, before payment of the supplier, should confirm that they comply with the requirements of the Contract, and that they are stored in such a way that no deterioration will occur. Investigations should also be made to make sure that no gap in insurance cover exists, while the materials are in storage, and later that the materials, together with all the precautions taken to prove ownership, remain in position.

Form of Tender (Appendix)
clause 60(2)

The maximum amount paid for materials is that percentage of their value stated in the Appendix to the Form of Tender, but the Engineer has some discretion to decide the proper amount (clause 60(2) (b)). To Engineers, value does not necessarily equate to cost, unless the cost appears to coincide with the material element of the relevant bill rate. Some Engineers are concerned that materials are brought on to site much too early and make a hefty discount on the amount certified for those materials to cover for any possible deterioration and extra labours needed to bring them up to standard. In such cases the Contractor can argue that there is an allowance in the rates for such contingencies or, possibly, that the period on site has been prolonged, by circumstances entitling him to recover any additional cost.

Retention

clause 60(4)

Each interim valuation is subject to a retention as described in clauses 60(4) and 60(5). The standard retention is 5% of the

clause 60(5)

valuation, excluding materials, with a maximum of 3% of the Tender Total except where the Tender Total is less than £50,000 when the maximum is £1500. The retention is reduced for each section or part of the project for which the Contractor has received a Certificate of Completion. The reduction is the lesser of 1.5% of the value of the section or part, or 50% of the retention held on the section or part.

clause 60(5) (a)

It helps the Contractor's cash flow to obtain as many of these partial Completion Certificates as possible. (clause 60(5) (a) and the notes on Completion Certificates, in Chapter 2). One half of the remaining retention is due for payment within 14 days of the issue of a Certificate of Completion for the whole of the Works, and the remaining half within 14 days after expiry of the last Period of Maintenance for the Works, if there are more than one.

clause 60(5)(b)
clause 60(5)(c)

clause 60(6)

However, the Employer is entitled to withold, from the second retention release, the Engineer's estimate of carrying out any outstanding work, until this work has been completed (clauses 60(5) (b) and (c)). In practice the Contractor can often find it difficult to obtain any of the second half of the retention money until maintenance has been completed. Some contractors tend to be slow to carry out maintenance work, which exacerbates the situation. Clause 60(6) entitles the Contractor to interest on any amount wrongly withheld, but this is probably second best to having the money.

It appears worthwhile to make an early start to any maintenance works to ensure these are completed by the end of the Maintenance Period. Non-completion can not only hold up release of retention but can also sour relationships, when these need to be maintained to reach agreements on the Final Account. Even if there are no claims to consider the Contractor should not allow the last memory of a job—and therefore the one most likely to be retained—to be one of niggling complaints over failure to pursue remedial work.

clause 60(5)

Clause 60(5) (a), (b) and (c) stipulates that releases of retention should be paid within 14 days of payment becoming due, which indicates that such payments are separate from interim certificates and are not affected by the minimum amount of these certificates.

clause 60(6)

Mention has been made of the Contractor's entitlement to receive interest on any money not paid by the time stipulated in the Contract. The amount of such interest, as stated in clause 60(6), is 2% above the minimum lending rate, which should cover the cost of borrowing money by a large company, but may not do so for a smaller one. The 2% is sometimes subject to amendment back to the one half of one percent, contained in the early versions of the ICE *Conditions*, Fifth Edition, which would not cover the cost of borrowing even by a large company. The problem is not always appreciated at site level, particularly where the contracting company does not charge interest to individual contracts. If interest is not charged to the site, late payments involving interest could enhance site results, and be seen as a benefit, when the delayed payment was to the detriment of the company as a whole.

Contract Price fluctuations

Contact Price fluctuations clause

The standard method of dealing with increases and decreases in prices, after the date of tender, which are to be reimbursed to the Contractor, is by means of the Contract Price fluctuations clause. This is not a part of the ICE *Conditions* booklet, and therefore is omitted from all Contracts, unless specifically included. At the time of writing, inflation is comparatively low, and the normal practice is to include the fluctuation clause in those contracts, with a contract period of more than two years.

The optional fluctuation clause is based on increases and decreases of indices, representative of the main resources used on civil engineering contracts, linked together by a formula, (often called the Baxter Formula, after the chairman of the committee that devised it). In fact there are two clauses because, in addition to that for general work, there is a special clause for use where the work is predominately fabricated steelwork. Both of these clauses are included as separate sheets with copies of the ICE *Conditions*. No more than an outline is given here of how the general clause works, with some notes on the advantages and disadvantages associated with its use.

The general method of adjustment for inflation–deflation is to multiply the value of work done during each month, 'the Effective Value', by an adjustment factor, which is the weighted average increase/decrease (from 42 days before the date for return of tenders to 42 days before the last day of the relevant valuation period), of a number of indices representing the most likely resources to be used on a civil engineering project. The indices that represent the value of resources at the time of tender are known as the base indices, Those that represent the costs for the month in which the work was done are the 'current indices'.

The resources are: labour, plant and nine materials, listed in section 4 of the clause. The proportions of each resource used in obtaining the weighted average increase–decrease are inserted by the Engineer, (in section 4 of the clause), before Tender. Together they add up to 90%, thus allowing a 10% non-fluctuating factor. The adjustment factor for a month is the summation of the product of proportion times the increase in each resource up to the valuation month). The proportions used to calculate the weighted average remain constant throughout the construction period, which leads to a divergence between actual increases–decreases in costs and those recovered through the formula, if the rates of change for each resource are not constant.

If the Contractor is paying suppliers or sub-contractors on a basis that more nearly represents their changes in cost, in times of inflation the Contractor will lose out on the higher than average inflation items—if concentrated in the later part of the Contract Period—but will have gained on those same items if executed early on. The opposite effect occurs to lower than average inflation items executed late or early in the Contract, because a preponderance of low inflation items executed in the early part of the Contract Period means that the percentage of higher inflation items will be greater later on and vice versa.

The indices, which are published by the Department of the Environment, are effective for a calendar month. It will be appreciated that a current index for a valuation period will be that for the previous month, if the period ends close to the end of a month (the current index being that which applied 42 days before the last day for which work is included, in the interim statement). If the end of a measurement period ends near the middle of a month, so that the 42nd day before it is just in the previous month, approximately the first half of the work done will be adjusted by the index that applies for that previous month, and the second half by that for the month in which the work was executed. In times of high inflation this is an advantage to the Contractor.

Problems have arisen in the past with regard to the indices' not reflecting the changes in cost e.g. at one time the index used for fuel was that for Derv rather than for gas oil the main fuel used on most civil engineering contracts, which had a much greater rate of inflation than Derv. Another example from the past was the availability of discounts on reinforcement, which were not taken into account in calculating indices, with the result that when they were withdrawn the effective price paid increased, but the index did not.

Because the formula is imperfect it is important to ensure that any arrangements for recovery of increases by suppliers and sub-contractors give away as little as possible. The obvious way of doing this is to have a back to back arrangement with all sub-contractors, whereby their recovery is based on the same formula as the Contractor's.

In the past some sub-contractors, particularly those laying asphalt paving, whose materials had a higher than average inflation rate, and whose work tended to occur at the end of the construction period, insisted on a formula with their own proportions of resources. The Contractor may be forced to accept this, but should at least insist that the proportions are not biased in favour of the highest inflating resource, which in the past has been coated roadstone. If it pays one sub-contractor to have his own formula, others will gain from sharing the main contract formula. The Contractor will probably lose less if all sub-contractors are either using their own formulae or all are taking a share of the main formula.

The Site agent should make sure he understands what has been assumed in the tender, with regard to any adjustments to the Tender to allow for variation between his recovery through a formula, and his actual increased–decreased cost. This is unlikely to be significant during periods of low inflation, but in the past adjustments have sometimes been quite large in relation to profit.

The Employer's right to make deductions from payments due

The Employer has a common law right to contra amounts due to him from the Contractor from payments to the Contractor. In addition to this, the Contract provides specific entitlement to make recovery, in a number of situations that could arise. These are

clause 25

(*a*) clause 25: the cost of premiums for insurance of the Works or Public Liability, paid by the Employer when the Contractor has failed to do so

clause 36(4)

(*b*) clause 36(4): the Employer's cost in carrying out inspections, tests, and so on which are the Contractor's liability but which he has not done

clause 39(2)

(*c*) clause 39(2): the Employer's cost of removal of materials or workmanship, during construction, which should have been removed by the Contractor, following an instruction by the Engineer so to do.

clause 49(4)

(*d*) clause 49(4): the Employer's cost of work to remedy workmanship or materials, discovered in the Maintenance Period to be not in accordance with the Contract

clause 53(8)

(*e*) clause 53(8): the Employer's unrecovered cost of selling or returning to the Contractor any of the Contractor's plant or materials not removed from the Site within the reasonable time allowed by the Engineer

clause 63(4)

(*f*) clause 63(4): the Employer's cost of having the Works

completed following the Contractor's expulsion from the Site for being in default.

Perhaps the most likely of these events to occur is that in (*c*) or (*d*) when the Contractor and Engineer are in dispute as to whether the work is, or is not, in accordance with the Contract. Even where there is a dispute there is no point in the Contractor's not carrying out the Work, when its removal becomes inevitable, if the cost of removal by others can be deducted from money that is due to the Contractor. The Contractor should make sure that he has recorded all the evidence to support his contention, before this evidence is removed.

clause 70

Value added tax—clause 70

Although included in this chapter, value added tax (VAT) is not really a part of the valuation, and is usually not dealt with by the Engineer. VAT is chargeable on the value of all taxable supplies to the Employer, by the Contractor, at either 0% or 15% of the value of the supply. These two rates might be changed, or more rates added. The Contractor also pays VAT —to sub-contractors on taxable supplies—but this does not affect the amount of VAT between Contractor and Employer (see chapter 9 under clause 19).

clause 19

The Contractor pays over to Customs and Excise any tax received (or rather tax that is chargeable) and recovers, from them, any tax paid out to suppliers and sub-contractors, on a netted-off basis. VAT is chargeable at the standard rate on the value of civil engineering work, but, with pressure from the European Commission to rationalise taxes throughout the EC, the situation may change in the future.

clause 70

The standard VAT clause in the ICE *Conditions*, clause 70, is on the basis that no VAT is included in any of the rates or prices, and gives the Contractor the right to obtain, from the Employer, any tax properly chargeable under the relevant Finance Act and VAT regulations. Paragraph (7) of the clause provides for a possible occurrence of a change in VAT tax status of the work done under the Contract, after the date for return of tenders. If such a change increases or decreases the cost to the Contractor of carrying out the Works, this is recoverable from or payable to the Employer. In the absence of the provisions of sub-clause 7 of the VAT clause the Contractor would have to absorb the cost of this tax. This is because, if some part of civil engineering work were made 'exempt', the Contractor would be unable to recover from the Customs and Excise any VAT paid to suppliers or sub-contractors.

As far as the Customs and Excise are concerned it is the

clause 70(7)

Contractor's responsibility to pay to them any tax due, whether or not it is collected from the Employer. It is therefore necessary for the Contractor to obtain a ruling from the local Customs and Excise office if there is any doubt as to the tax status of any of the work done. Clause 70(5) deals with the procedure to be adopted if there is any dispute between Contractor and Employer, as to the proper amount of VAT chargeable, or if the Employer disagrees with the opinion obtained from the local VAT office.

clause 70(5)

clause 70

Clause 70 opts for the supply by the Employer of an 'Authenticated Receipt' as the Tax Document instead of requiring the Contractor to issue a Tax Invoice. It is important not to include an actual amount of VAT on the Contractor's valuation as this might be regarded as making the valuation statement a Tax Invoice, thus bringing forward the date when tax would have to be paid to the Customs and Excise.

Final account

clause 60(3)

The Contractor is required by clause 60(3) to submit a statement of final account, backed up by supporting documentation, not later than three months after the date of the Maintenance Certificate. There is nothing in the ICE *Conditions* that prevents the Contractor from submitting the statement at an earlier date, and in general it is to his advantage to make the submission as early as possible. There is also nothing in the ICE *Conditions* to state what happens if the submission is delayed beyond the three month period, but the Contractor will no doubt be involved in some unnecessary arguments and delay if he misses the deadline. The best policy is to avoid unnecessary arguments, wherever possible, and concentrate on those that are essential.

The final account will be the compilation of all the amounts of work done, based on billed or varied rates or prices, including preliminaries plus all other matters claimed, on the basis of additional cost, mentioned earlier in this chapter.

6. Claims

The clauses to which reference is made in this chapter are: 52(4) for procedure and 5, 7(3), 12, 13(3), 14(6), 17, 26(1), 27(6), 30(3), 31(2), 32, 36(2), 36(3), 38(2), 40(1), 42(1), 50, 52(1), 52(2), 52(3), 55(2), 56(2), 60(6), 69, 71, for events giving rise to claims. Other clauses mentioned are: 2(3), 20(2), 22(1), 51, 50(1).

The ICE *Conditions* are one of the few sets of standard conditions to use the word 'claim' in the text, although the word is commonly used in connection with most types of contract, during their execution. The common use of the word 'claim' is in connection with disagreements between Contractor and Engineer or Architect or Employer, but, as used in the ICE *Conditions*, it not only covers situations where a disagreement has occurred, but also others where the claim is the first notification, to the Engineer, of an occurrence which may or may not lead to a difference. Claims mentioned in the ICE *Conditions* are as follows:

clause 44(1)

(a) the contractor's request for an extension of time is referred to as '..claim to extension of time..' in clause 44(1) of the conditions, although there may be no difference of opinion as to entitlement. Clause 44 deals with extension of time only, and not payment for the cost of the delay that gives rise to that extension.

clause 52(1)
clause 52(2)
clause 56(2)
clause 52(4) (a)
clause 55(2)

(b) if the Contractor does not accept a rate or price determined and notified to him by the Engineer, pursuant to clause 52(1), 52(2) and 56(2), he claims a higher rate or price, under clause 52(4) (a). Rates notified under clause 52(1) and (2), in addition to those for variations, include those in connection with the rectification of errors in, and omissions from the Bill of Quantities (clause 55(2)). These are claims in the broader sense as they arise following a statement of the Engineer, with which the Contractor disagrees.

clause 52(1)
clause 52(2)
clause 52(4)

(c) claims for additional payments, other than those under clause 52(1) and (2), must be notified, in writing, to the Engineer, under clause 52(4) (b). These claims range from the cost of samples (clause 36(2)) and additional labour taxes (clause 69)—neither of which should prove to be difficult to

clause 36(2)
clause 69
clause 7(3)

agree on—to additional cost caused by the Engineer's delay in the issue of drawings and other information (clause 7(3)). If there is a personal involvement on the part of the Engineer's staff who are dealing with the latter claims, they can become controversial, both in principle and evaluation.

clause 56(2)
clause 52(4) (a)
clause 52(4) (b)

It would appear that claims for higher rates under clause 56(2) need to be notified under both clauses 52(4) (a) and (b) but it is suggested that this is due to an error in clause 52(4) (b). It seems reasonable to assume that the intention, in drafting the two separate sub-clauses, was to differentiate between claims based on rates and those based on cost. This chapter concentrates on the monetary claims of categories (*b*) and (*c*), and others which fit into neither of these categories. Claims for extensions of time that fall within category (a) above are covered in chapter 3.

clause 13(3)
clause 14(6)
clause 31(2)
clause 52(4) (b)
clause 69

It is doubly clear from such clauses as 13(3), 14(6) and 31(2), that notification of the claim must be made under clause 52(4) (b), because the clause contains a reference to this requirement. However, the wording of clause 52(4)(b) is comprehensive enough to include notification of claims for additional payments made under other clauses, which contain no reference to clause 52(4) (b), and the Contractor should act accordingly. A claim for additional costs, due to changes in level or incidence of taxes, under clause 69, is an example.

clause 13(1)
clause 66(1)

The Engineer is endowed with a much wider authority to issue instructions and give decisions regarding disputes between the Contractor and Employer, under the ICE *Conditions*, than is the case under other standard conditions of contract. The wide power to issue instructions is contained in clause 13(1), which a prudent Engineer tends to ignore, and that for decisions comes from clause 66(1), which states

> 'If a dispute or difference of any kind whatsoever shall arise between the Employer and the Contractor in connection with or arising out of the Contract ...'.

clause 22(1)
clause 22(1)(b)

This appears to be broad enough to allow the Contractor to ask the Engineer for his decision on breach of contract claims, or responsibility for damage to third party property, where the Employer argues that the damage is covered by the general indemnity of clause 22(1) and the Contractor considers that one of the exceptions in clause 22(1) (b) applies to the situation. The Engineer might also be requested to give a decision on any disagreement over the amount due for repair of damage caused by one of the excepted risks, or for dealing with archaeological remains, both of which are executed 'at the expense clause 20(2) clause 32 clause 52(4)

of the Employer' (clauses 20(2) and 32).

There is a reasonable argument that none of the situations mentioned in the last paragraph need be notified under clause 52(4). However, prompt notice should be given of all potential problems.

One reasonably common claim on large contracts, for which there is provision in some conditions of contract, but not the ICE *Conditions*, is that known as an acceleration claim. Some comment on the negotiation of acceleration between Contractor and Employer has been made in chapter 3, but a claim can arise where the Engineer rejects a claim for extension resulting from the Employer's delay, to which the Contractor believes he is entitled. In these circumstances, limitations of the time available severely reduces the efficacy of arbitration as a means of settling the dispute on extension. Therefore the Contractor is faced with the choice of accelerating to complete in time, or taking the extra time for completion, to which he considers he is entitled. Whatever decision he makes he will need to prove his entitlement to extension, at some stage, to establish a monetary claim for acceleration or for the extra time spent on site. Acceleration claims are dealt with in greater detail below.

clause 52(4)(a) clause 51 clause 56(2) clause 55(2)

Claims arising under clause 52(4) (a)

These are claims for rates greater than those fixed by the Engineer, pursuant to a variation issued under clause 51, change in quantities complying with clause 56(2), or a disagreement on rates arising from an error in, or omission from, the Bills of Quantities in accordance with clause 55(2). The steps that lead to a claim are

clause 51 clause 51(2) clause 52(2) clause 56(2)

(*a*) the Engineer or Engineer's Representative (ER) issues a variation order under clause 51, in writing; *or* the Contractor confirms a verbal order in accordance with clause 51(2); *or* the Contractor or Engineer give notice that a variation has an indirect effect on the rates or prices for other items of work. Notice has to be given before the varied work commences or, as soon thereafter as is reasonable (clause 52(2)); *or* a change of quantity from that given for the item in the Bill of Quantities—which affects the rate for that item or any other items of work—becomes apparent (clause 56(2))

(*b*) the Engineer (or ER) and the Contractor consult with regard to the rates or prices to be applied to the affected items of work (clauses 52(1) and (2) and 56(2))

clause 52(1) clause 52(2) clause 56(2)

(*c*) the Engineer (or ER), in the absence of agreement with the Contractor, determines the rates and prices and notifies the Contractor (clauses 52(1), 52(2) and 56(2))

clause 52(1)
clause 52(2)
clause 56(2)
clause 52(4) (a)

(*d*) if the Contractor does not accept some or all of the rates of prices, he must notify the Engineer (or ER) in writing, within 28 days of the notification by the Engineer of the determined rates or prices, that he intends to claim higher rates or prices (clause 52(4) (a)).

There is an additional step to incorporate in this procedure if the Engineer's Representative has determined the rates or prices, acting under delegated powers. The Contractor has the right to refer the matter to the Engineer for his decision (clause 2(3)). This action would be taken in lieu of step (*d*). However, should the Engineer confirm his representative's rates or prices, the Contractor should be prepared to make a claim under clause 52(4) (a) within 28 days of that confirmation.

clause 2(3)

clause 52(4) (a)

The Site agent, in asking the Engineer to review a decision of the Engineer's Representative, should be aware that the Engineer is going to rely on his Representative for any detailed analysis of the situation. To succeed in persuading the Engineer to alter a decision, the Site agent must be able to isolate some key points or facts that the Engineer can consider, rather than submerge them in detail.

There is a fixed, maximum time of 28 days, after receiving notice of fixed rates or prices, in which the Contractor must give notification of his intention to claim. It is therefore important to know when the Engineer considers that rates have been fixed. There may be a number of communications from the Engineer, setting out his rates or prices, during the consultation stage, including copies of interim statements with the Contractor's rates changed. To avoid recriminations and unnecessary arguments, an agreed method of denoting the fixing of a rate would be advisable. The 28 days' period, within which the Contractor is required to give notice of his claim, is important, but failure to keep to this timing is not fatal to the claim. In accordance with clause 52(4) (e) the Contractor remains entitled to payment to the extent that the Engineer has not been prevented from, or substantially prejudiced in, investigating the claim. Nevertheless prompt notification of claims by the Contractor allows consideration to be given to the merits of the principle and valuation of the claim, without wasting precious time on arguing whether or not the investigation of the claim has been prejudiced by a delay.

clause 52(4)(e)

clause 52(4)(b)

Claims arising under clause 52(4) (b)

Claims, other than those for rates and prices, do not go through any preliminary stages but become claims straight away. These claims are based on those clauses, listed in chapter 5 on pages 94 to 96, which entitle the Contractor to additional payment based on cost, should the specified occurrence arise.

clause 52(4)(b)

Before the Contractor has any entitlement to extra payment, over that based on the work done at bill rates, he must give the Engineer notice of his intention to claim, under clause 52(4) (b), as soon as reasonably possible after the occurrence of the event that gave rise to the claim.

Other claims

clause 52(4)
clause 52(4)(a)
clause 52(4)(b)

The procedure and timing, prescribed in clause 52(4), applies only to those claims for which notification has to be given under clause 52(4) (a) and (b). Breach of Contract claims, for instance, remain open during the relevant Statute of Limitations period. However, delays in notification, by either party to the Contract, can bring with them problems for the adequate proving of the principle or valuation, and should be avoided.

Claims procedure after notification

clause 52(4)

Once notification has been given under clause 52(4) (a) or (b) the procedure for dealing with both value and cost based claims is more or less the same—i.e.

(*a*) for claims notified under clause 52(4) (b): 'Upon the happening of such events the Contractor shall keep such contemporary records as may reasonably be necessary to support any claim he may subsequently wish to make.' The events referred to are those that give rise to the claim (clause 52(4) (b) and chapter 5, pp. 94–96)

clause 52(4)(b)

(*b*) the Contractor must keep any contemporary records, or further such records, as are reasonable and may be material to the claim, that the Engineer instructs should be kept (clause 52(4) (c))

(*c*) the Engineer must be allowed to inspect all records kept in support of a claim, and be given such copies as he may request (clause 52(4) (c))

(*d*) the Contractor should, as soon as is reasonable after notification, submit a first interim account giving full details of grounds for, and amount of, the claim to that date (clause 52(4) (d))

(*e*) the Contractor should continue submissions of details of the claim, at such intervals as the Engineer may require (clause 52(4) (d))

(*f*) the Engineer shall certify in interim certificates and the Employer shall pay so much of the claim as the Engineer shall consider has been substantiated by the submitted details (clause 52(4) (f)).

Step (*a*) means that the Contractor has to start keeping records, for claims submitted under clause 52(4) (b), on the occurrence of the event that gives rise to the claim, which is before he is required to submit notice of his claim. This does not apply to claims based on rates submitted under clause 52(4) (a), presumably because these are expected to involve adjustments of Bill rates and prices, and will have been discussed, as variations, before they became claims.

clause 52(4)(b) clause 52(4)(a)

If the Contractor's failure to comply with any of the requirements of clause 52(4) prejudices the Engineer's investigation of any part of the claim, the Contractor loses entitlement to be paid for that part (clause 52(4) (e)). The records that the Contractor should maintain, without instruction (clause 52(4) (b)), appear to be the same as those he can be instructed to keep (clause 52(4) (c)). If the Contractor keeps the required records for cost claims, it might be supposed that instructions on further records should be unnecessary. However, there are differences of opinion as to what records are reasonably necessary to support a claim. In the absence of guidance from the Engineer, the Contractor should inform the Engineer of the records being kept, when the claim is notified. This should put the onus on the Engineer to decide if he considers these to be sufficient, and avoid later arguments.

clause 52(4)(b) clause 52(4)(c)

It is not uncommon for the Engineer to require more records than the Contractor considers necessary, and for the Contractor to put too much emphasis on the value and too little on the principle on which the claim is based. A criticism of Engineers, by Contractors, is that they have an insatiable desire for more and more records, rather than for taking a decision in principle, based on the information in their possession. The Contractor who submerges his arguments on principle within the valuation has less cause for complaint, in this respect.

The Contractor should be aware that unless he can demonstrate to either the Engineer's or, failing that, an arbitrator's satisfaction, that he is entitled to some additional payment, he will receive nothing. This applies to both the reason for being paid at all, and also the amount of the payment. Proof of losses is not, in itself, proof of entitlement to extra payment. The summing up of a claim once seen, which not surprisingly proved to be optimistic, was: 'we are not able to prove that we did it but, on the other hand, the Engineer cannot prove that we did not, so, in the end, we must receive at least 50% of the claim.' Proof of the principle is important, but of course the keeping of records to verify the amount cannot be left until agreement on principle has been reached. It is only where careful thought is given to the situation that claim situations are recognised in time for the appropriate records to be maintained.

Sometimes the pro rata of a rate affected by a variation, or the additional cost caused by some event, may require records to have been kept before the claim arose, so as to provide a 'before and after' comparison. Examples under the clauses of the Contract that give entitlement to claims, as follow, indicate the circumstances.

Claims on rates and prices

The agreement of a varied rate is more likely to become a claim if the new rate is considerably in excess of that anticipated by the Engineer. For instance a change from sand to pea gravel in the filling of open jointed drains might be expected to require a higher rate. However, if the site is so isolated that the Contractor has to supply his own materials, the change could be from the use of a waste material to one that involves the production of more waste.

Take the case of a temporary sand and gravel pit opened to supply concrete aggregates, on a large scale, which contains a large surplus of sand which can be used around open jointed pipes. While the material specified is sand, it is available from what otherwise would be waste material, but if the specification is changed to pea gravel, this would need to be obtained by extra excavations in the pit. If the pea gravel was 10% of the material in the ground, the change in specification would entail the handling the remaining 90% of material, and also the rehandling of the surplus sand, back into the pit. The price of pea gravel filter would bear no relation to that for sand, when priced on the above basis. Good records would be needed to substantiate the rate for the material and that it could not have been purchased more cheaply. There is a danger that the Contractor, having decided initially that site-produced materials were more economic than those purchased from outside sources, would produce filter material without making a further assessment for pea gravel. This situation could result in the Contractor pressing for an unnecessarily high price. However, if the high price is correct but is notified only after completion of the work on drains, the Engineer may become aggrieved because, had he been aware of the price in time, he might have withdrawn the variation order.

Problems arise in pricing work if the method of measurement being used involves the use of theoretical quantities, such as the assumption, in earthworks, that 1 m^3 of excavation produces 1 m^3 of fill for an embankment. If, under that assumption, a job has 1 million m^3 of excavation and an embankment of 1·1 million m^3, then 100 000 m^3 of imported material should be billed, and paid for. However, if the excavation compacts by 10%, the filling imported would be 200 000 m^3, whereas if it bulked by 10%, no imported filling would be used. In either case pricing of the bill items would need to take into account the relation between the amounts paid for and those used.

Contractors could choose to place any adjustment for this relationship in a variety of ways in the Bill of Quantities. If a 10% compaction was anticipated this could be accommodated by doubling the rate for imported fill; allowing a lump sum in the preliminaries or spreading the amount of the adjustment over all the earthworks rates. Any change in quantity or variation which affects the relationship or the quantity of imported filling could cause a problem for the adjustment of rates. However, if the Contractor put the adjustment on the right items, he could end up satisfied, without the need for a claim. There would also be a problem in fixing a rate for imported filling if no import had been required and this item had been omitted from the Bill in error. The Contractor would be entitled to have the error rectified, by the inclusion of an item for imported fill, but the rate would have to be a fair rate. It is not always easy to persuade Engineers that an item should be added to the valuation, when no work has been done (clauses 55(2) and 52(1)).

General claims

The general claims which occur frequently on contracts are

(*a*) for reimbursement of additional oncosts and overheads for extra time spent on part of, a section of or the whole of the Works caused by extra work or delays that the Employer's responsibility, the time may be within, or outside the Times for Completion in the Contract

(*b*) for delay and disruption to work items owing to extra work or delays that are the responsibility of the Employer.

A third which tends to appear from time to time is the 'Acceleration Claim' which arises when the Contractor considers he is entitled to an extension, which is not granted. He claims the extra expense of providing additional resources to complete in time, but less efficiently than if he had taken the extended time to which he was entitled.

Additional oncosts and overheads

The claim is often referred to as 'additional preliminaries' or 'valuation of extension of time', but an extension of time is not essential to a successful claim to recover the expense of additional time. The Contract requires the Works to be completed within the Times for Completion, so that an event could cause additional, recoverable expense, even where the extra time required still allows completion within the then current completion time (see the comment on Fig. 3 in chapter 3). The Contractor should bear in mind that not all extensions of time entitle him to additional payment. This depends on the cause of the delay.

The Contractor's entitlement to recover the oncosts and overheads can arise in three ways, and the method of valuation depends on the cause of the extra time, as follows.

(*a*) If the time arises as a result of variations or changes in the quantities of work stated in the Bill of Quantities, the Contractor is entitled to adjustment of any rates or prices made unreasonable by the changes. These would include those rates and prices affected that contained allowances for oncosts and overheads (clauses 52(2) and 56(2)).

clause 52(2)
clause 56(2)

(*b*) If the cause of the extra time is a delay due to an event other than variations or changes of quantities, for which the Employer is deemed to be responsible, the Contractor is able to claim the cost from the Employer, under the relevant clause. A list of the clauses that allow the Contractor to recover cost plus overheads or cost plus overheads plus profit, is given in chapter 5, pages 94 to 96

(*c*) If the extra time is caused by a breach of Contract on the part of the Employer, the Contractor is entitled to claim damages for that breach.

Adjustment of rates for Site oncosts

There are no set rules for adjusting rates and prices under clauses 52(2) or 56(2). It is left to the Engineer:

clause 52(2)
clause 56(2)
clause 52(2)

'the Engineer shall fix such rate or price as in the circumstances he shall think reasonable and proper' (clause 52(2)) and 'the Engineer shall after consultation with the Contractor determine an appropriate increase or decrease of any rates or prices rendered unreasonable or inapplicable ...' (clause 56(2)).

clause 56(2)

In practice, it is more usual for the adjustment for extra time to be calculated as a total amount and paid as a lump sum, rather than follow the Conditions of Contract fully, and allocate that sum as adjustments to all affected rates and prices.

The Contractor must expect the Engineer to make an assessment of any additional, time-related overheads—recovered by way of the extra quantities of work included in interim certificates—to set against those needed for the extra time. For instance if all the preliminaries are spread equally over the work items, a 10% increase in value will give a recovery of 10% more preliminaries. On a broad brush basis, if the changes cause a 10% increase in time, there is no need for adjustment to the rates. In contrast, if all the preliminary costs were included in a 'General Items Bill' then, on the same broad basis, all the time-related general items would need to be increased by 10%, if the time was increased by that amount.

In practice the preliminaries are likely to be partly in a general items bill and partly spread on the rates, but not necessarily equally on all the rates. The spread on the rates may have been uneven, possibly because the Contractor priced for enhanced early cash flow, or he had a policy of adding oncosts and so on to labour and plant, materials and subcontractors, at different percentages. Not all time-related preliminaries cover the whole of the time needed for the execution of the Works, and adjustment for these would require individual treatment.

In order to adjust rates and prices the Engineer will need to know exactly how preliminaries are allocated to bill items and varied items, and a detailed analysis of the make-up of the costs allowed. He may also need to satisfy himself that the figures given him by the Contractor are reasonable—e.g. that having taken away the money for preliminaries, there is sufficient left in the production items to execute the work.

Broadly speaking the non-productive costs fall into three categories

(*a*) fixed costs such as the setting up and removal of site offices, plant, temporary roads and other temporary works
(*b*) costs that vary with time such as staff, maintenance of offices and temporary roads, site transport, standing charges for heating, lighting, water and telephones, hire of scaffolding, tower cranes, pumps and other general plant, unloading gangs, chainmen, cleaners, and watchmen; by no means all of these costs are spread over the full time of the contract—some apply to specific

structures, operations or parts of the project only, and may fluctuate in level over the period in question.

(*c*) costs that vary with the value of the work done such as insurance of the Works, public liability and employer's liability insurance, national insurance, protective clothing and small tools.

The division of costs into these three categories is an approximation only. No item can be considered to be confined to any one category, whatever variations take place. A variation that increases the weight of units to be lifted by tower crane may involve the replacement of the anticipated machine by a larger one, thus increasing the fixed element and also the weekly charge for its hire, before the time adjustment is made.

The final premium for insurance of the Works is related to the final value of the Works, but the risk of damage is increased if it occurs over a longer period, and thus the premium could be increased for this additional risk. Third party and public liability insurances, protective clothing, and other items are related to costs of wages and salaries rather than the cost of the work.

The cost of electricity, telephone calls, gas and water would often be taken as a weekly charge, varying perhaps with the time of year, but there could be a case where telephone calls and water were related to the volume of work being done at a particular time. The number of telephone calls might increase where large numbers of variations are involved.

Although the Contractor may wish to evaluate his claims for the expense of extra time by the use of simple pro ratas, he may find, on many occasions, that it is advantageous to make more complex calculations, or the Engineer may not accept that the broad brush approach allows for swings and roundabouts. Having set out all the time-based preliminary items it may be necessary to allocate each to the period in the Contract, when the costs were to be incurred, and then connect the additional time for each item to the event causing prolongation. The necessity for this type of calculation increases with the number of separate parts there are to the project.

Figure 6 shows the bar chart approach to anticipated staff on site. A similar drawing could show those necessary as a result of changes. The Engineer should be in a position to check that the numbers of staff on site before any variations fitted the numbers anticipated. This sort of illustration can be produced on a computer using spread sheets, adding values of each resource for each week.

Most contracts are based on fixed prices, so the prices in the

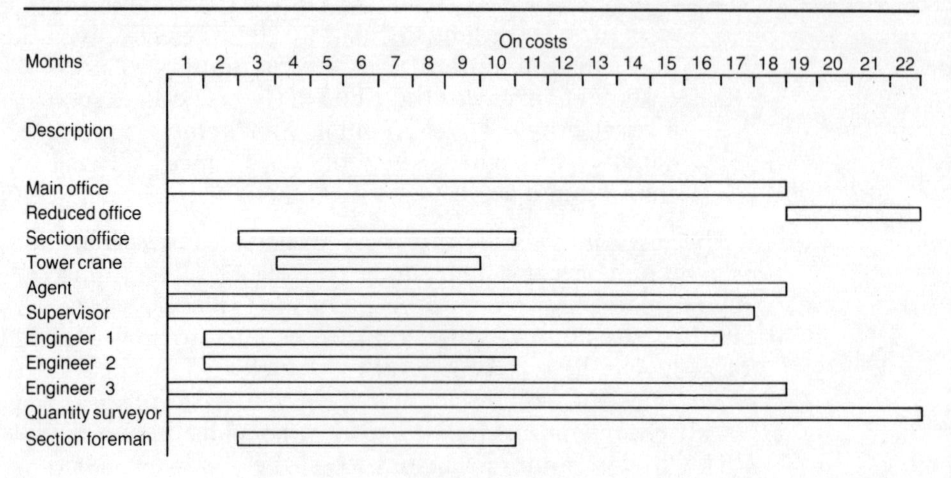

Fig. 6. Bar chart for site on-cost items

Bill for oncosts will be the weighted average over the period in which the work is done. The additional preliminaries will need some adjustment for inflation, or for deflation should that ever occur. The inflation amount cannot be calculated by adding some percentage to the additional preliminaries claimed, because the main effect may be the transfer of all the original costs to a later time in the Contract period.

Adjustment of rates for off-Site overheads
The adjustment of rates to cover for off-Site overheads can cause problems, partly because many of the staffing and other costs of the Contractor's off-Site offices may not be involved in the contract in the additional time period. The other part of the problem is that overheads are often expressed as a percentage of value, which gives rise to the belief that there is no case for extra overheads, except where value increases.

The Contractor has a certain cost per annum for his total overhead costs, which, apart from inflation, will be fixed, unless the Contractor is expanding turnover. A proportion of this fixed amount, based on the anticipated turnover per year of each particular tender project compared with the total anticipated turnover, is included in that Tender. If the Contractor obtains and executes his planned amount of work, he will recover all his overheads, but if not, recovery will be more or less than cost.

Some contractors may charge actual overheads to projects, as a percentage of actual turnover, but others may charge on the tendered percentage thereby taking any profit or loss on overheads at head office level. Whichever method of accounting is used, it is unusual to penalise a project that has not achieved its anticipated turnover, and is the cause of a loss on overheads. Neither is the project which has an increase in turnover and is responsible for a profit on overheads, rewarded. Take for example a contract, expected to run for two years, which runs for three years, without an increase in value. If overheads are accounted for on the basis of actual turnover, the amount recovered in the two-year period will be only two thirds of that expected to be recovered. If the prolongation is the Contractor's responsibility, then he grins and bears it, but if it is deemed to be that of the Employer, under a clause in the Contract that allows reimbursement for the effects of the delay, the Contractor would claim additional overheads. If the prolongation is caused by some event which entitles the Contractor to adjustment of the rates and prices, he would seek compensation for the unrecovered amount caused by the variation or increased quantities.

The broad brush approach would be

$$\text{additional overheads} = \frac{\text{overheads in tender}}{\text{original time}} \times \text{prolongation}$$

This calculation gives no credit for any additional overheads recovered for increased quantity of work executed, but a separate adjustment could be made. The formula gives a reasonably correct answer only for a project with one line of activities, executed at an even pace, but totally inaccurate if the delay occurs in only one of several sequences of activities, or during a period of reduced resource level on Site, which allowed some resources to be released for work on other sites. The calculation can be adapted by using several formulae on each sequence of operations, but a method of allocating overheads, based on the cost of on-Site staff, might be better received by the Engineer.

The theory is that the Contractor's ability to carry out construction work depends on the availability of staff, to supervise and administer that work. If staff control possible turnover, the most reasonable way to allocate overheads is as a percentage of the cost of the staff needed on site. Recovery of additional head office overheads for prolongation would be based on the percentage of the amount for site-based supervision in the claim. This might be more acceptable to the Engineer, because it is probably similar to the method he uses to allocate his general costs. A credit for the extra recovery of overheads resulting from the increased value of work done, would be made if the percentage was applied to the net extra cost of site staff, priced at the allowances in the Bill of Quantities.

Financing charges

One particular charge which may be included in general overheads, or may be charged directly to Site by some Contractors, is the cost of financing the project. There is now a wide acceptance in the industry that any effect on the financing charge should be taken into account, in the adjustment of rates and prices affected by variations or changes of quantities.

The Contractor will be required to demonstrate the amount and location of this cost, in the Tender, and how, and to what extent, it has been affected by variations or changes in quantities. On a broad basis the contractor's costs arise and are paid for in the following manner

- labour is still mainly paid for on a weekly basis, usually on the Thursday following the week in which the work was done
- materials costs arise one month and are paid for at the end of the following month, but there may be considerable exceptions to this general rule
- sub-contract costs arising one month are due to be paid 42 days after the end of the measurement period in accordance with the FCEC Sub-contract Conditions, but there may be exceptions
- staff costs arising in one month are paid at the end of that month, but there is a tendency towards payment in the middle of the month
- plant and equipment costs follow materials costs, as far as outside hire is concerned, but each contractor has his own system for internal plant
- there are other costs such as electricity, water, national insurance, insurances, rates and taxes, which follow other patterns.

clause 60(1)

The income against these costs arises from the interim payments, which depend on how quickly the Contractor submits his statements, under clause 60(1).(If these are sent within a few days of the end of the valuation period, the Contractor should receive payment around the end of that month, but these may not reflect, exactly, the costs for the same period. Additional financing cost to the Contractor arises if the procedure followed to obtain payment, or full payment for variations, prevents the values being included in the certificate for the month in which the work was done.

Separate calculations for each individual variation create an unnecessary amount of work, and it may be possible to agree an average figure for all variations. The Contractor would be asked to demonstrate that any charge for financing has been removed from other overhead calculations. It is normal to treat as separate any calculation of additional finance charge caused by the holding of retentions over extended periods.

Where the Contract includes the application of the fluctuation formula to the payments due, any later inclusion of variations into the certificates will involve their adjustment by a later—and probably higher—fluctuation factor than otherwise would have applied. This often has been accepted as making up for any compensation due to for the extra finance charges, although the two adjustments have no contractual connection.

Payment for extra time on a cost basis

If the delays causing additional oncosts and overheads are caused by one of the events that entitle the Contractor to any cost caused by the event (see chapter 5, pages 94 to 96) the additional resources can be determined in the same way as if they were caused by variations, but the amount due is based on actual cost rather than on tender allowances. As with variations there is the need to demonstrate that the amounts claimed were caused by events giving entitlement to reimbursement.

Disruption claims

Disruption claims arise from the same events which give rise to extra oncost and overhead claims, but recovery is sought for the reduced output on the affected operations. Disruption should also lead to extra time, unless additional resources can be brought on to compensate for the reduction in output. The extra time might relate to a part or a section of the Works only, but could affect the completion of the whole of the Works.

Reimbursement to the Contractor for disruption is made either by adjustments of the rates (if caused by variations or changes in quantities) or by payment of additional cost if caused by other events allowed by the ICE *Conditions*. Individual instances can easily be illustrated in principle, but convincing quantitative proof of the valuation is often more difficult. Sometimes the effect of any change can be deduced from calculations, in other cases the disruption can be established from outputs, before and following the events claimed to have caused the reduction.

The most difficult disruption to prove is that claimed to have

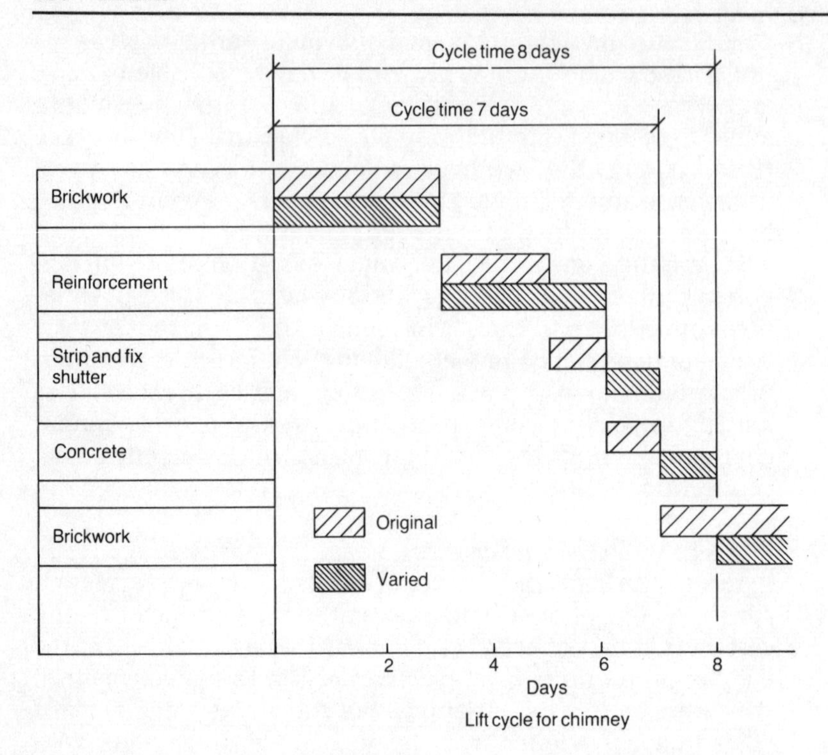

Fig. 7. Original and varied bar chart programmes for one lift of concrete chimney

occurred from the beginning of construction, resulting from a combination of variations and other causes, particularly lack of detail. These involve proof that tender outputs could have been achieved but for the problems introduced by the Employer–Engineer, and the removal of any disruption cost (deemed to be the responsibility of the Contractor) from the sums claimed.

A simple example is illustrated in Fig. 7 which shows the programme and progress for a reinforced concrete chimney that is being constructed in lifts. The amount of reinforcement is increased, by a variation which, because of the limited working room available, results in a need for additional time for reinforcement fixing. As a consequence the cycle time for each lift is extended and, if no other work is available, the efficiency of the remainder of the team working on the lift is reduced. The adjustment to the rates for brickwork, shuttering and concreting can be made by multiplying the allowances for labour and plant for these items in the Tender by the new and then dividing by the anticipated cycle time. The overall time for the chimney will also be prolonged, which will necessitate adjustment of any rates that contain an element of time-related preliminaries. The issue may be clouded if the programme times for brickwork, shuttering and concrete are shown by events to have been optimistic, but the pro ratas would need to be made using bill rate allowances for labour and plant, but actual timings.

Another example, allowing a calculation to be made of the new rates, arises from the ordering of additional compaction of the fill on an excavation and filling operation. If the variation resulted in the overall output of the team being limited to the amount of material that could be compacted, when previously the team had been balanced, the efficiency of both the excavation and filling operations would be reduced. A calculation of the efficiency of excavating and filling, based on the old and new number of passes, could be backed up by a further calculation based on the outputs before and after the change in compaction requirements. If possible the Contractor should bring in additional compaction plant, but this is not always easy to find at short notice in the middle of a muck-shifting season. The formula is

$$\text{efficiency} = \frac{\text{original number of passes}}{\text{new number of passes}} \times 100\%$$

The rate would need adjustment by multiplying it by the reciprocal of the efficiency.

A similar loss of efficiency might occur to the bulk excavation and filling exercise if drainage blankets of imported free draining material were introduced as a variation. If the change to the design occurred after some filling had been completed it might be possible to compare outputs before and after the introduction of the drainage blankets. In this case there is no simple theoretical calculation for the efficiency of the changed operation.

If records could be kept of the changed operations, these could be taken to differentiate between workings close to each side of the blanket and those on the section of earthworks midway between two drainage layers. The possible effect of the variation on progress has to be realised quickly, if records that will highlight the effect satisfactorily are to be organised. If no records are possible or available there can be a comparsion only between the output assumed in the Tender and that achieved, with some comparison between the Tender quantity per hour assumed, and the maximum potential output of the plant. This could provide a reasonable indication that the anticipated rate of working should have been achieved, had the variation not been issued. With all records of earthworks operations it is necessary to have complete records of breakdowns and weather conditions, in order to make reliable comparisons of outputs.

Additional winter working is a special case of disruption, which often features as a separate claim heading. The basis of the claim is that events deemed to be the Employer's responsibility, or additional work, caused the deferment of weather-susceptible work—programmed to be carried out in the summer—to the winter months. The reason for the delay to the work has to be established, by the Contractor, as a precedent to any recovery.

There are general effects such as a greater incidence of stoppages owing to inclement weather, the changed ratio between working time and paid time as a result of the shorter day and the more difficult general effect of working in less congenial conditions to assess. The decrease in efficiency is often accompanied by a requirement for additional resources, such as additional haul roads for earthworks, heating of aggregates and water and the protection of placed concrete. Contractors are apt to forget that the forced movement of work into winter is often accompanied by movement of other operations into better conditions.

Individual causes of disruption with a consequent loss of efficiency can be relatively simple to demonstrate, but the overall effect of a number of events on the efficiency of the work as a whole, often referred to as the knock-on effect, is much more difficult to prove to the Engineer's satisfaction.

This overall disruption claim can be highly exaggerated, and often is, even though based on the sound principle that many changes make control of the project more difficult and some breakdown of that control more likely. A loss of control, leading to a general reduction in output, arises if the Contractor is prevented from introducing or maintaining an incentive bonus scheme for operatives. If a scheme is to be effective, it requires not only the willingness and ability—on the part of the labour force—to exceed the targets set, but also the availability of a sufficient and regular supply of materials and work to allow the bonus to be earned. If the work is not available, bonuses cannot be earned, and pressure is put on the site management to pay fixed bonuses, or those operatives capable of earning the large bonuses leave to try somewhere else (bonus schemes are considered briefly in chapter 2).

Take for example a scheme where the operatives are paid 100% of the saving in time achieved and a 30% bonus can be earned. For every 13 h paid the Employees will do 10 h work. If the scheme is a true incentive scheme, that 10 h work would require 13 h, without the incentive. Assuming the operatives would require the same weekly bonus, if a fixed bonus were paid, they would work 13 h and be paid for $13 \times 13/10 = 16{\cdot}9$ h, to achieve the same output for which they would be paid 13 h with the incentive scheme. The additional labour cost to the Contractor, on a broad basis would be 30%. This is the theoretical additional cost of a perfectly working scheme, which is impossible to achieve in practice, over all operations, even if the Engineer and Employer do nothing which interferes with the potential of the scheme. Other adjustments might have to be made, and other incentive schemes would produce different results, but the broad calculation could be on similar lines to the example given. Where an operation involved plant, the saving caused by an incentive bonus scheme would include a reduction in time and cost for the plant.

The amount of output that might be lost with the loss or partial loss of an incentive bonus scheme can be considerable, but the problem of gaining the Engineer's acceptance that this has been caused by circumstances that entitle the Contractor to reimbursement is difficult to surmount. This is not likely to succeed, if thrown in at the end of construction, as support for a disruption claim. To stand any chance of success representations would need to be made to the Engineer, as events occur, to demonstrate the difficulties encountered in the meeting of targets, to allow the Engineer to see for himself the extent of the problem. If the Contractor can show that he is endeavouring to overcome the problems, but failing through no fault of his own, it should be possible to convince a reasonable Engineer. However, sympathy may be easier to come by than certification.

Another broad basis calculation for general disruption can be based on the involvement of additional time involved, which causes reimbursible expense. If the same amount of work is spread over a longer period, without the opportunity to reduce labour or plant resources, the additional percentage cost of labour and plant is approximately equal to the percentage of the additional time taken. An adjustment to the calculation could be made if extra work, of a smaller percentage than that for additional time, was involved. This broad brush applies to the situation where the work is affected by a large number of changes, spread out over the period, and for a single sequence of activities. A number of calculations could be made, on the same basis, if several activity sequences were affected in the same way. The Contractor would be required to prove that the extra time was caused by events that entitled recovery of extra cost or adjustment of the affected rates, and that resources could not be reduced.

If the period of overall disruption follows a period in which the Contractor has not had to cope with significant variations, it may be possible to illustrate the changed conditions through a diagram, such as that given in Fig. 8. Both the recovery from the valuation and the cost of labour and plant resources are plotted against time. The timing of each variation is indicated by an asterisk plotted against the relevant time. This type of illustration should indicate that the separation between the value and cost lines follows an increase in numbers of variation instructions. It will surprise no one this occurs in Fig. 8. The unrecovered resources applied is indicated by the difference between the two lines. In the case shown, the two lines for value and cost more or less coincide initially, indicating that the Contractor is able to cope until large numbers of variations occur. There may be other factors which explain the divergence, such as unequal pricing in the Tender, or coincidental delays by the Contractor. If no other reasons are apparent, it is not unreasonable to assume that the loss of production was caused by the weight of variations.

Delays which entitle the Contractor to reimbursement of any cost caused can be dealt with in the same way as claims for the effect of variations and changes of quantities, but the valuation of these is based on cost rather than on rates and prices. Under the various clauses in the ICE *Conditions* listed in chapter 5, the Contractor is entitled to the costs that stem from these events only, which does not equate, necessarily, to the total cost of the operation less the amount recovered through the rates and prices. This equation applies only if the allowances in the Bill of Quantities' rates and prices were adequate and the total cost contains no non-recoverable cost.

Acceleration claims

There is no provision in the ICE *Conditions* that allows the Engineer or Employer to order the Contractor to complete in times shorter than the Time for Completion for Sections or the Whole of the Works, stated in the Form of Tender (Appendix), adjusted for any extensions of time to which the Contractor is entitled.

Form of Tender (Appendix)

If the civil contract is part of a much larger project such as a chemical processing plant, then the cost to the Employer of delays could be very large, in comparison with the value of the civil contract (see the section on Liquidated Damages in chapter 3 and examples in Appendix 4). Where the Contractor is entitled to an extension for those delays and the Employer

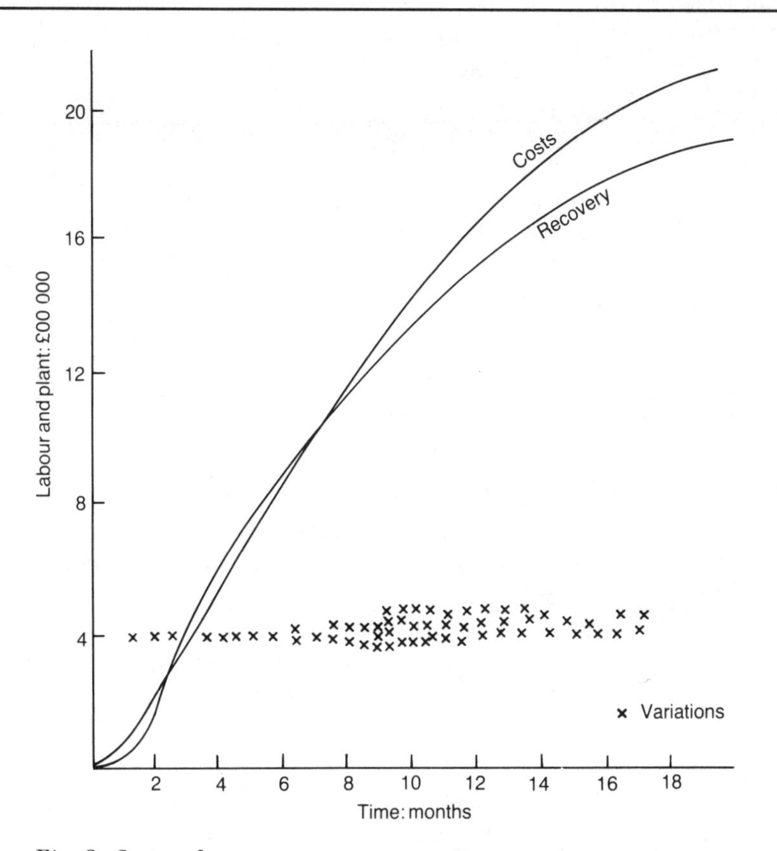

Fig. 8. Cost and recovery against time illustrating effects of increasing numbers of variations

receives no compensation for the costs of the delay from other sources, he may prefer the Contractor to complete without recourse to the extended time. The answer is for the Employer to try to negotiate an acceleration agreement with the Contractor, which, in effect, is the amendment of the time provisions contained in the Contract. The Engineer may be authorised by the Employer to carry out these negotiations on his behalf, but the Engineer is not authorised to do so under the Contract.

The chances of achieving any acceleration depend on making an almost immediate decision, in order to leave the maximum time available for the accelerated programme, which leaves no room for cat and mouse negotiation techniques. The Employer will want an agreement which guarantees that the acceleration will be achieved or, in the event that it is not, that damages will apply. On the other hand, the Contractor might want the payment for acceleration to apply for the effort he makes regardless of its success.

The negotiation is a commercial one, based on the advantages that accrue to each party from the change to the Contract, and should not be constrained by the prices in the Contract, or the cost of the measures taken by the Contractor. Neither party can force the other to accept any proposal put forward and, if no agreement is reached, all the terms of the Contract continue to apply to the completion of the Works.

If the Employer is hoping to save a million pounds from the acceleration, he is more likely to obtain a quick response from the Contractor if he is willing to pay a fair proportion of that to pay for the cost and any extra risk taken by the Contractor. An agreement is probable only if both sides put their cards on the table from the outset.

Any agreement reached should cover all aspects clearly, and the two simple methods of doing this are

- the acceleration payment is paid to the Contractor in compensation for waiving his rights to a specified extension of time entitlement
- the acceleration payment is made to the Contractor as compensation for reducing the Contract period by the amount of time equal to the existing extension of time entitlement; the extension is then awarded, thus restoring the original completion date.

Both of these methods leave all of the other original contract provisions in place.

Circumstances may arise where the Employer would want to reach some acceleration agreement, even though the delay was the Contractor's responsibility. This situation might arise where the Liquidated Damages figure in the Contract was considerably lower than the anticipated damages. It is reasonable to assume that in such cases, the chances of success are not good, particularly as, up to that time, the Contractor had been unable to cope with the progress requirements of the Contract. However, there is the odd occasion where the expense of retrieving lost time may be more expensive to the Contractor than paying for Liquidated Damages. The negotiations described should not end up as claims unless it had been agreed that acceleration was to be attempted, but consideration for doing so had not. This sort of situation should be avoided, as a number of other difficulties might arise, such as what would happen if none of the time, or only part of it, were saved.

The situation most likely to give rise to an acceleration claim arises when the Contractor claims an extension of time, which, in the opinion of the Contractor, the Engineer wrongly rejects, in part or whole. The Contractor, if convinced he is correct in his belief, could dispute the Engineer's award and complete the works in the time to which he believes he is entitled. This would involve payment of Liquidated Damages, if the Engineer maintained his decision on extension, until it was overturned after recourse to arbitration. However, the Contractor, while disputing the Engineer's decision, could decide to take measures to complete in time, and submit an acceleration claim. One or other of these decisions have to be chosen because it is virtually impossible to resolve such a dispute quickly enough, through arbitration, for the decision to be implemented in completing the Works.

If the Contractor assumes he is correct with regard to extension entitlement, his decision will depend on the particular circumstances that apply to the situation. If the Employer saves a considerable sum by the earlier completion, he is likely to be better pleased with the Contractor who takes the acceleration option. At the same time if the acceleration is achieved, there should be money available to pay the Contractor's expenses. If the Contractor is of the opinion that acceleration is unlikely, then there seems little point in attempting it, as the costs of trying to save time but failing are unlikely to be recovered.

Between these two extremes there will no doubt be some difficult choices to be made, and the Contractor should also look at his decision on the supposition that he is incorrect in his assessment of extension entitlement. If the Contractor goes for acceleration the situation will not be the same as when the entitlement to an extension has been accepted by the Engineer, with the possible exception of where the Engineer's decision not to make an award has been influenced by the Employer.

If the principle of the claim is established, the valuation is likely to be on the basis of the additional cost of acceleration compared with completion in an extended time. Some credit allowance against extra costs would be expected where the extension entitlement was in respect of some event, such as exceptionally adverse weather conditions, for which no entitlement to reimbursement accrues for the additional costs of retaining Site organisation for the extended period.

Additional costs of acceleration include

- increased staffing and support costs to deal with the higher work intensity
- importation of extra plant, possibly at a higher hire rate on a short hire basis or because the plant available has a greater capacity than that required

- additional payment to operatives to encourage recruitment of the larger work force requirement
- additional overtime payments
- loss of efficiency owing to long hours, congested conditions at work stations, reduction of time intervals between successive operations causing greater disruption
- increase in material costs owing to importation from other suppliers, or to greater wastage
- claims from suppliers and sub-contractors
- additional temporary works or materials owing to adopting concurrent working on two or more operations, in lieu of sequential construction.

Disputes

clause 66

Clause 66 is often referred to as the arbitration clause but it is rather more than that. It also deals with the necessary preliminary action that has to be taken to open the way to arbitration. The initial procedure is for either party to the Contract —i.e. Employer or Contractor—to refer any dispute, in writing, to the Engineer for his decision. It is only after such decision has been given, or if no decision is given in a period of one or three calender months of the referral (depending on whether a Certificate of Completion for the whole of the Works has or has not been issued) that reference to an arbitrator can be made. Resort to the initial procedure may settle the matter, but it would be imprudent for a Contractor to assume that the referral of a dispute to the Engineer, under clause 66, will cause sufficient consternation to bring forth a prompt and acceptable decision.

clause 66

The matters that can be put to the Engineer for his decision, under clause 66, cover an extremely wide range, much wider than for other Conditions of Contract as can be verified from the excerpt from the following clause 66(1)

> 'If a dispute or difference of any kind whatsoever shall arise between the Employer and the Contractor in connection with or arising out of the Contract or the carrying out of the Works including any dispute as to any decision opinion instruction direction certificate or valuation of the Engineer (whether during the progress of the Works or after their completion and whether before or after the determination abandonment or breach of the Contract) it shall be referred in writing to and be settled by the Engineer who shall give his decision in writing and give notice of the same to the Employer and the Contractor.'

This includes matters which the Engineer has the authority to decide, under some other clause of the Contract, but does not exclude items on which the Contract is otherwise silent, or indeed, breaches of contract. Examples of these items are dispute as to whether damage to third party property was the Contractor's responsibility or came within one of the exceptions given in clause 22(1) (b), or the proper payment due to the Contractor, for repairing damage to the Works caused by one of the 'Excepted Risks'.

clause 22(1)(b)

If the decision which the Contractor disputes was given by the Engineer's Representative, acting under delegated powers, then this should first be referred to the Engineer for confirmation or change (clause 2(4)), before being submitted to him under clause 66.

clause 2(4)
clause 66

The Site agent should not initiate any of the procedure leading to arbitration, without instruction from his head office, so it is important not to do so inadvertently. If the Engineer has given a clear final decision with regard to some matter, he might regard a further letter to him on the subject as a reference under clause 66, unless it contained points that had not been raised previously. In addition reference to the Engineer of matters such as responsibility for third party damage might also be regarded as coming under clause 66, because the Contract gives the Engineer no power to make a decision on these matters, other than under clause 66. The Engineer might be wrong in his assumption, but time could be wasted in unnecessary argument.

clause 66
clause 66

The Site agent is completely safe, in this respect, in writing to the Engineer's Representative on matters being dealt with under delegated authority, because both should be aware that a reference, under clause 66, can only be made to the Engineer (clause 2(3)).

clause 2(3)

7. Determination before completion

The clauses to which reference is made in this chapter are: 2(3), 3, 4, 13(1), 15(1), 16, 39(2), 40, 41, 43, 46, 51(1), 51(2), 63, 64 and 65.

General

This chapter deals briefly with the possibility that the contract will be left uncompleted, or completed by someone other than the Contractor. It is to be hoped that experiences of this type of situation are very rare, and if events do arise, no doubt a decision will be made in the Contractor's head office, with the aid of legal advice, rather than on the Site.

Nevertheless, the actions of the Site staff could bring a situation to a head, where expulsion of the Contractor from the Site is contemplated. Some awareness of the possible consequences of certain actions may ensure that the Site agent does not become an unwitting victim of his own actions.

Legal or physical impossibility

Under the Contract, the Contractor is required to construct, complete and maintain the Works, in strict accordance with the Contract, unless it is legally or physically impossible to do so (clause 13(1)).

The first part of the proviso confirms the obvious—that a Contract to do something illegal cannot be enforced under the law. One example is the absence of planning permission, possibly because some significant variation to the Contract involves a further application for planning permission. It would not be prudent for the Site agent to continue working on the assumption that planning permission will catch up with events on the ground, without obtaining advice. Although the Employer warrants that all planning permissions have been or will be obtained (clause 26(2) (c)), he may not be financially capable of fulfilling that warranty.

clause 26(2)

The second proviso should not cause a problem with the completion of the Contract, in most cases. In the UK, if some part of the contract is physically impossible to construct, a variation can be issued, under clause 51, to avoid the problem area.

clause 51

Any impossible aspect of the Engineer's design can be amended by variation, if the impossible nature is recognised, and provided the project does not lose viability as a result of heavy additional costs of the variation. The Employer would be in default if he abandoned the Contract in such circumstances. If the whole of the project becomes impossible owing to something catastrophic, like the Site's being washed away, no doubt that would constitute 'Frustration' which is the subject of clause 64. Unfortunately for the Contractor, 'physically impossible' does not equate to prohibitively expensive, so there is no relief for the Contractor who has woefully underestimated the cost of completion.

clause 64

Suspension longer than 3 months plus 28 days

clause 40

Under clause 40, the Engineer is empowered to give an order, in writing, that suspends all or a part of the Works. However, if a suspension lasts 3 months, the Contractor may, unless the suspension is provided for in the Contract or is a result of his default, ask permission, in writing, to proceed with that part of the Works to which the suspension order refers. In the event that permission is not received within 28 days of the receipt of the request, the Contractor, by a further written notice, may elect to consider the suspended part as not required by the Employer (clause 40(2)).

clause 40(2)

clause 51

clause 52(1)

clause 52(2)

If the suspension is for a part only of the remaining work, then it is treated as an omission, under clause 51. The values of any omission and its effect on other rates in the Contract are ascertained in accordance with clause 52(1) and (2) respectively (see chapter 5). If the suspension order affected the whole of the remainder of the Works, failure to lift it, following the Contractor's request, allows the Contractor to assume that the Employer has abandoned the Contract, putting him in breach of Contract.

Although the Contract contains the remedy for a number of breaches, there is none in the conditions for this particular default. In England, the Contractor would be entitled to the same financial outcome he would have enjoyed had the breach not occurred (i.e. he would finish with the profit he would have made after all costs had been taken into account). The overall reimbursement could be based on one of two methods

(*a*) if little work had been carried out before the abandonment, it would be simpler to base the value on cost, including that involved in the acceptance of special materials ordered or the cost of abandoning orders for materials, plant or sub-contract work and the reasonable cost of having staff and operatives await redeployment, plus a loss of profit.

(*b*) where a great deal of work had been completed then the Bill of Quantities could be used to value this with adjustments, if necessary for changes in quantities, plus cost for materials and commitments plus a loss of profit on outstanding work.

In both cases the problem area might be the amount of cost by way of Site and off-site overheads that should be allowed in the valuation. The problem is more acute in those situations where only a small part of the Works was completed before abandonment.

Forfeiture following the Contractor's default

clause 63

Clause 63 deals with the action to be taken by the Employer, if the Contractor becomes insolvent or has assigned the Contract without the Employer's consent, or the Engineer has certified, in writing, that the Contractor has committed one of the offences listed in clause 63(1) (a)–(e)

clause 63(1) (a)–(e)

(*a*) has abandoned the Contract; or

(*b*) without reasonable excuse has failed to commence the Works in accordance with clause 41 or has suspended the progress of the Works for 14 days after receiving from the Engineer written notice to proceed; or

(*c*) has failed to remove goods or materials from the Site or to pull down and replace work for 14 days after receiving from the Engineer written notice that the said goods materials or work have been condemned and rejected by the Engineer; or

(*d*) despite previous warning by the Engineer in writing is failing to proceed with the Works with due diligence or is otherwise persistently or fundamentally in breach of his obligations under the Contract; or

(*e*) has to the detriment of good workmanship or in defiance of the Engineer's instructions to the contrary sub-let any part of the Contract.

With such an occurrence the Employer, after giving 7 days' notice in writing to the Contractor, can expel the Contractor from the Site, and complete the Works with his own resources or those of another contractor, at the Contractor's expense. He can also recover his own additional costs from the Contractor, provided that there is enough money in the kitty or the Contractor is not insolvent.

The Site agent will have little control over the solvency of the Contractor, or the assignment of the Contract. However, he may be able to prevent his head office from informing the Employer that the Contract has been transferred to a different part of the group, due to internal reorganisation, without first obtaining the Employer's permission to do so (clause 3). It is not suggested that this would cause the Employer to take action under clause 63, but it just could upset relations sufficiently to ensure permission to assign was refused, when it was eventually requested.

clause 3

clause 63

However, the offences listed in clause 63(1) (a)–(e) may well be within the Site agent's control, and he should be aware of the possible consequences. In practice there appears to be little difference between (a) and (b) as there must be some reason to assume that the Contractor has abandoned the Contract. Apart from informing the Employer or Engineer of his intention to abandon, the most likely indication of abandonment is failure to start or to continue work, without a reasonable excuse. There can be a difference of opinion on whether or not the Contractor should close down operations over the winter, especially if the work is behind programme at the time. This situation may very occasionally boil over, so the strength of the Employer's–Engineer's convictions should be carefully gauged.

clause 63(1) (a)–(e)

Failure to remove goods, materials or workmanship which are not in accordance with the Contract, after being informed by the Engineer, in writing, that they have been condemned and rejected, is another situation which occasionally involves brinkmanship. Contractors are loathe to remove work, unless it is well out of specification, but, without the agreement of the Employer, the Engineer has no discretion to accept work not in accordance with the Contract. The Engineer would not be expected to resort to issuing a certificate to the Employer under clause 63, except in the most serious of instances of failure to remove condemned work, but the possibility that he will exists.

clause 63

The problem is more acute in cases where other work is to be built on work that has not been approved by the Engineer, because time is important if delayed completion is to be avoided. It would be imprudent of the Contractor to build on something that was not approved, and to delay demolition, in the hope of a change of mind on the part of the Engineer, may only result in further additional cost becoming involved in the equation.

Other items, standing on their own, might be subject to a change of mind. However, the Contractor should take into account the effect on the end date if no change occurs. The risk of adding Liquidated Damages to the overall cost might outweigh the possibility of a change on the part of the Engineer. As time passes and the total of Liquidated Damages increases, the possibility of the Engineer changing his stance must diminish. (see chapter 2).

The Contractor needs to review the situation with his advisors before the Engineer has issued a certificate under clause 63, and this requires the Site agent to be able to judge the situation, from contact with the Engineer or his Representative on Site. Although only the Engineer can issue a certificate, (clause 2(3)) the Engineer may well be obtaining the information, on which his decision will be based, from his Representative on Site.

clause 63

clause 2(3)

Where the Contractor disputes the Engineer's condemnation of a certain item, but the Engineer insists that he is correct, the Contractor needs to ensure that he has all the necessary proof to support his case before demolition. Even where the Engineer has given a certificate under clause 63(1) the Employer need not take action to expel the Contractor from the site. Clause 39(2) allows the Employer to have improper Work and/or materials removed by others at the Contractor's expense. One would expect action to be taken under clause 39(2), except in the most extreme cases, where the Employer/Engineer considers that action under clause 63(1) is the only way to obtain the workmanship required with a minimum of delay.

clause 63(1)

clause 39(2)

The offence in clause 63(1) (d) is the Contractor's persistent or fundamental breach of his obligations under the Contract. The one example given in the clause is the failure, despite previous warning in writing, to proceed with the Works with due diligence. The Contractor would be in breach of his obligations under clause 41, although this clause uses different wording. The Engineer's notification under clause 46, or a written reminder to the Contractor that he was not complying with his obligations, under clause 41, could constitute warnings. The added threat of a clause 63 notice could force the Contractor to accelerate progress to complete in time, where he and the Engineer disagreed on the amount of extension of time that should be awarded (see acceleration claims in chapter 6).

clause 63(1)(d)

clause 41

clause 46

clause 41

clause 63

It is suggested that the fact that the Contractor is not working at a pace that will ensure completion within the Contract Period does not, in itself, entitle the Engineer to issue a certificate under clause 63. From a practical viewpoint, if progress is as swift as is possible in the circumstances, following a warning, it would seem that the Engineer ought to accept

clause 63

that the Contractor is proceeding with due diligence. The Engineer, in such a situation, would have difficulty in demonstrating that the Employer's best interests would be served by bringing in a new company, to complete the Works.

clause 63(1)(d)

Other breaches coming under clause 63(1) (d) could include failure on the part of the Contractor to provide superintendence and operatives of the necessary experience and skill (clauses 15(1) and 16) and refusal to remove any person who, in the opinion of the Engineer, has misconducted himself or is incompetent or negligent (clause 16). It is suggested that in this case the Engineer would require supporting evidence for his opinion.

clause 15(1)
clause 16

clause 63(1)(e)

The last breach mentioned is in clause 63(1) (e) and is the sub-letting of work without the Engineer's prior written consent (clause 4). However, the breach of clause 4 must either be to the detriment of good workmanship or be in defiance of the Engineer's instructions not to sub-let.

clause 4

In the event of forfeiture the Contractor is not entitled to any further payment until the end of the Period of Maintenance. Any payment then is subject to the total cost to the Employer of having the work completed and maintained being less than the Contractor's entitlement had he completed everything himself. The chances that further payment might be due to the Contractor are slim, if the change of contractors has involved the charging of Liquidated Damages against the Contractor as a consequence of delay in completion.

clause 63

Where the Employer does take action, under clause 63, the Contractor is entitled to expect that the total cost to complete including the cost of delay should be kept to a minimum. However, the ICE *Conditions* do not give the Contractor the right to take part in the ascertainment of the value of the Works, completed at the time of forfeiture, or what would have been the Contractor's Final Account.

Frustration

Many contractors' personnel use the word 'frustration' in connection with their claims for extra payment, in the sense that they have been prevented, by the Employer, from working in the way that had been planned. Thus a disruption claim would be referred to as a 'Claim for Frustration', which may correctly describe the Contractor's feelings, but could be misinterpreted contractually.

clause 64

The word 'frustration' should be reserved for use in connection with the events mentioned in clause 64—i.e. where the Contract has been stopped short by events outside the influence of either of the parties to the Contract, such as acts of government, or extremes of storms, making the continuation of the Contract impossible.

War clause

In the case of war, in which the UK is involved, the Contractor is required to continue to execute the Works in accordance with the Contract, as far as it is possible to do so, for a period of 28 days—at the end of which period, the Employer is entitled to determine the Contract. On determination the Contractor is required to remove plant and equipment from the Site, as soon as possible, and he is entitled to the final valuation set out in clause 65(5). Briefly this consists of Work and preliminaries completed at Bill rates and prices, together with a proper proportion of those rates and prices for items partially completed, the cost of materials reasonably delivered to Site or which the Contractor is legally liable to accept, plus the amount of any expenditure reasonably incurred in the expectation of completing the Works, and in the removal from the Site.

clause 65(5)

Contractor's right to determine the Contract

There is no clause in the Contract that sets out the Contractor's remedy where the Employer is in persistent or fundamental breach of the Contract, but he has such a remedy at common law. If such a breach occurs, he is entitled to cease the execution of the Works, when satisfied that he is leaving the Works in a safe state, and would be entitled to damages for the breach. These damages would probably be similar to the valuation on determination following an outbreak of war, but with the addition of compensation for loss of profit.

Termination by agreement

It is always open for the parties to the Contract to agree to cancel a Contract by agreement, the terms of such agreement depending on how much each side want to be relieved of their responsibilities and liabilities under the Contract.

8. Sub-contractors

The clauses to which reference is made in this chapter are: 4, 7(3), 8(2), 21, 22, 24, 31(2), 35, 44, 52(4) (b), 53, 58, 59A, 59B, 60, 61, 63, 64, 65, and the Form of Sub-contract published by the Federation of Civil Engineering Contractors (FCEC Sub-contract).

Domestic Sub-contractors

Very few projects are carried out now without any of the work being sub-contracted, although there is still less sub-contracting in civil engineering than in building, particularly in relation to Sub-contractors nominated by the Employer.

clause 4

The general attitude in the Contract is set out in clause 4. Sub-letting the whole of the Works is forbidden, and the Contractor must obtain the written permission of the Engineer, before sub-letting any part, unless he is required to do so under the terms of the Contract.

clause 58(5)

Sub-contractors whom the Contractor is required to employ, under the terms of the Contract, are referred to as Nominated Sub-contractors, as defined in clause 58(5). The Contractor's own choice of sub-contractors have no other contractual name, but are often referred to as 'Domestic sub-contractors', and this term is used in this book, if it is intended to exclude Nominated Sub-contractors from the reference.

clause 58(5)

clause 4

In accordance with clause 58(5), Nominated Sub-contractors can be either sub-contractors or suppliers, but it is thought that the restrictions on sub-letting in clause 4 apply only to domestic sub-contractors, and not to material suppliers. Clause 4 makes it clear that the provision of labour on a piecework basis does not constitute sub-letting and, although not stated, neither would the hiring of plant on a time basis. However, a plant hirer taking on the excavation for a rate per unit, or for a lump sum, would be a sub-contractor.

clause 4

Failure to obtain the Engineer's written consent to sub-letting, or going ahead with sub-letting, after the Engineer has given a written refusal to a request, can result in the Contractor's expulsion from the Site by the Employer (clause 63(1) and chapter 7). Clause 4 does not require the Contractor to obtain obtain the Engineer's consent to a specific sub-contractor, but rather to the act of sub-letting, to anyone. That assertion is not universally accepted by Engineers and, apart from that, the Contract documents often contain a requirement to name domestic sub-contractors.

Probably the best way to obtain consent is to write asking for consent in principle to sub-let the required activities, and give the names of the successful sub-contractors as soon as they are known. If the Engineer objects to any of those chosen, the Contractor can raise the matter of interpretation of clause 4, after hearing the reasons for the objection, if he wishes to persist with that sub-contractor. It is a good idea to avoid arguments on principle, before it is known, in practice, that the principle will come into play.

clause 4

Clause 4 requires that the Contractor will remain as responsible to the Employer for the sub-contractor's performance as he is for his own, and this responsibility is in no way abated by the Engineer's having consented to the sub-letting.

Nominated Sub-contractors

Nominated Sub-contractors are suppliers or sub-contractors with whom the Contractor is required to enter into an agreement to supply goods, materials or services (clause 58(5)). The requirement can be expressed by

clause 58(5)

(*a*) the supplier or sub-contractor's being named, at the time of Tender, in a Prime Cost Item in the Bill of Quantities, or possibly elsewhere in the Contract, but with reference to a particular item (clause 58(5))

(*b*) the issue by the Engineer of an instruction with regard to a Prime Cost Item in the Bill of Quantities (clause 58(4))

(*c*) as in (*b*) but the instruction is in respect of a Provisional Sum (clause 58(3)).

clause 58(5)
clause 58(4)
clause 58(3)

Provisional Sum and Prime Cost (PC) Item are defined in clauses 58(1) and 58(2) respectively.

clause 58(1)
clause 58(2)

Objection to a nomination

The Contractor has the right to object to any Nominated Sub-contractor on reasonable grounds, and to any who decline to enter into a sub-contract containing the provisions set out in clause 59(1) (a)–(d). These provisions put the Sub-contractor in the same position vis-a-vis the Contractor as that existing between Contractor and Employer, but they are not the only conditions to be considered to ensure a true back-to-back arrangement.

clause 59A

For those Nominated Sub-contractors named in the tender

documentation any objections, on reasonable grounds, need to be considered at the time of Tender. If, for instance, only the name is given in the Prime Cost Item, objections on the grounds of the financial status of the nominated company, or based on previous experience of very bad workmanship and so on, could be made with the Tender. The usual ban on submission of a qualified tender may make a real objection difficult, but some reference to the need for discussing nominations could be made.

If full details of the sub-contractor's tender are supplied with the tender documentation, the Engineer might expect other objections to be raised at that time, particularly those which might not be made by all tenderers, such as the construction time required by the sub-contractor.

Some grounds for objection to nominations, that may arise are

- the time period for his work, required by the Sub-contractor, is, in the Contractor's opinion, too long in relation to a section or the whole of the main contract work
- the sub-contractor expects the Contractor to supply some service or equipment which is not provided for in the Contract
- the Nominated Sub-contractor's offer to the Employer contains some design element which is not expressly provided for, under clause 58(3), in the Contract (see also clause 8(2) and chapter 2 under 'Contractor's Liability for Design')

clause 58(3)
clause 8(2)

- the time interval required by the sub-contractor before he can start work on Site is too long to ensure completion of a section, or the whole of the Works, by the relevant date for completion; where this is a result of a delay by the Engineer in issuing the nomination instructions, it may justify a Contractor's claim for an Extension of Time and costs (clauses 7(3), 44, and 52(4) (b)); however, the Contractor should make the objection

clause 7(3)
clause 44
clause 52(4)

- a sub-contractor for building type work, e.g. asphalt waterproofing, has tendered on the basis of measurement on the Building Method of Measurement, in lieu of the ICE method (CESMM)
- the Nominated Sub-contractor insists on the incorporation of Liquidated Damages in the sub-contract, rather than being responsible for any damage he causes the Contractor.

Any objections that the Contractor has with regard to a nomination should be made before he places an order with the proposed sub-contractor. Once an order has been placed, a contract exists, which limits the action that can be taken by the Engineer under clause 59A(2). In all probability, the Engineer will be very unsympathetic, particularly if he would have been able to remedy the situation by nominating another sub-contractor, had the sub-contract order not been placed. The Contractor should be wary of making objections that cannot be sustained, for any time lost in the execution of the Works as a result may have to be recovered at his cost.

clause 59A(2)

Some time and trouble might be saved, on any nominations still to be made, if the Contractor offered to vet the invitations to tender, before these were sent out. To accept the offer, the Engineer would need to respect the fact that this was made in a spirit of co-operation and helpfulness, with no wish to criticise, or to take responsibility away from the Engineer.

The actions that the Engineer can take on receipt of an objection to a nomination are listed in clause 59A(2). The nomination of another Sub-contractor could apply in the case of any of the objections already listed (clause 59A(2) (a)).

clause 59A(2)

clause 59A(2)

A variation of the Works, including the omission of the sub-contract supply (clause 59A(2) (b)) could also be used to overcome all the examples of objections, but there are some limitations. The Engineer cannot insist on a variation which added the Sub-contractor's design into the Contract. This can only be done with the Contractor's agreement, but the more contractual method of dealing with the problem is to take the design out of the sub-contract. The Contractor would then be made responsible to the Employer for the design, or possibly, to the Engineer, if it was part of the design he had undertaken. The solution to the problem is in the hands of the Employer, Engineer and the proposed sub-contractor.

The work of a proposed, Nominated Sub-contractor can be omitted from the Contract, and executed under a direct contract with the Employer, if the Contractor raises a valid objection (clause 59A(2) (b)). This course of action would not be effective in all cases—for instance, where the Contractor's objection was that the proposed sub-contractor required more time than could be accommodated in the Contract Period. A change to a direct subcontractor would be ineffective, unless he could complete in a shorter period than that required by the proposed nomination. The Contractor would just change the clauses under which he made his claims. However, use of a direct contractor would be suitable, if the Contractor objected on the basis of past experience of very poor workmanship from the nominated company.

clause 59A(2)

Where the work is omitted the risk to the Contractor must be reduced but he would remain entitled to his percentage, for clause 59A(2)

charges and profit, for the omitted work (clause 59A(2) (b)).

If the Engineer, with the consent of the Employer, requires the Contractor to enter into a Nominated Sub-contract where the terms are not consistent with those of the Contract, then the Contractor is not bound to discharge those obligations and liabilities which are inconsistent (clause 59A(2) (c) and 59A(3) (a)). At first glance this solution to an objection seems to be the way of solving any problem, but on examination this is shown not to be the case. Take for example an objection on the grounds that the Sub-contractor insisted on the inclusion of Liquidated Damages in the Sub-contract, which were a value based pro rata of those in the Contract. The Contractor would be liable to the Employer for that fraction of damages only, when delays were caused by the sub-contractor. These damages would be recovered from the Sub-contractor, but would leave the Contractor suffering his own costs of the delay, and that of any other Sub-contractors affected.

clause 59A(2)
clause 59A(3)

The last action open to the Engineer, to overcome an objection, is to instruct the Contractor to carry out the work, which was to be sub-contracted (clause 59A(2) (d)). However, as is stated in clause 58(4), this cannot be done without the consent of the Contractor, and would not be a viable alternative if the nominated company had a patent on the process to be used, or the prospective sub-contractor had a sole rights agreement with the manufacturer to erect the specified materials.

clause 59A(2)
clause 58(4)

Employer's responsibility for Nominated Sub-contractors

The Contractor is as responsible for the work of Nominated Sub-contractors as he is for the work he does himself, or that is executed by domestic sub-contractors (clause 59A(4)). Nevertheless, the Employer does undertake to waive certain rights to recover costs from the Contractor that arise from the default of a Nominated Sub-contractor, if the money cannot be recovered from the sub-contractor (clause 59A(4)). To avoid having to pay the Employer the Contractor must do all he can, as required by the Employer, to enforce the Sub-contract or obtain other remedies available to him, to recover any amounts due from the Sub-contractor. The steps to be taken include proceedings at arbitration or in the courts, but with the Employer paying the Contractor the reasonable costs of any action (clause 59B(6)).

clause 59A(4)

clause 59B(6)

If a Sub-contractor is in breach of the Sub-contract, which causes delay to the Contract, then the Contractor suffers his costs of delay, in addition to being liable to the Employer for Liquidated Damages. If the amounts are large enough, and the possibility of recovery from the Sub-contractor is good, it would be reasonable to cover both points in any proceedings. In such a case the Employer and Contractor each take a fair proportion of the costs and expenses of the action, in so far as these are not recovered from the Sub-contractor (clause 59B(6)).

clause 59B(6)

If a Nominated Sub-contractor's breach of Sub-contract is serious enough to entitle the Contractor to apply the forfeiture clause in the Sub-contract, or to consider that the Sub-contractor has repudiated the Sub-contract, he should inform the Engineer, in writing, to obtain the Employer's consent to taking action (clause 59B(2)).

clause 59B(2)

However, the Contractor is not penalised, in any way, if he takes action without consent, provided he has the right to do so and follows the correct procedure when taking the action (clause 59B(4)). The point here is that the Contractor is entitled to have any delay consequent on the forfeiture taken into account for an extension, and to be reimbursed any additional cost, necessarily and properly incurred as a result of the delay. A delay before the actual forfeiture might not be considered consequent on the forfeiture. The Contractor must keep the Engineer informed of his own costs, if permission to terminate is requested and refused. Presumably a refusal would count as an instruction under clause 13(1) to continue using a defaulting Sub-contractor, and claims would be made under clause 13(3).

clause 59B(4)

clause 13(1)

clause 13(3)

If a Sub-contract is terminated the Engineer is required to take one of the choices he makes following an objection to a nomination, except he cannot continue with the defaulting Sub-contractor.

Sub-contract conditions

The Federation of Civil Engineering Contractor's Form of Sub-contract (FCEC Form) was specially prepared for use for sub-contracts where the main contract is under the ICE *Conditions*. As far as it is possible to do so, the sub-contract form refers back to the ICE *Conditions*, and hence will be usable even where some amendments have been incorporated into the main contract. However, care should be taken to make sure that any such amendments need no change to the sub-contract conditions, and the sub-contractor is made aware of the changes.

The FCEC Form has been approved by the Committee of Associations of Specialist Engineering Sub-contractors and the Federation of Associations of Specialists and Sub-contractors. Thus members of these associations should not object to the use of the conditions, but sometimes do, particularly with reference to damages and payment.

The ICE *Conditions* stipulate that certain provisions should be included in any sub-contract, and this is done by the FCEC Form, which is suitable for both nominated and domestic sub-contracts.

The provisions are as follows.

clause 53

- The requirements of clause 53, regarding the plant, temporary works, equipment and materials brought on to Site, and the vesting of the property in them in the Employer, is included in clause 11(1) and (2) of the FCEC Form.

FCEC clause 11(1)
FCEC clause 11(2)
clause 63(2)
FCEC clause 2(3)
clause 16(1)

- In accordance with clause 63(2) the Employer may require a sub-contract to be assigned to him, in the situation where the Contract has been forfeited owing to the Contractor's default. This has not been specifically provided for in the FCEC Form but the Contractor is not barred from assigning the sub-contract, under clause 2(3) of this form, and the sub-contract is not automatically determined on the determination of the Contract (clause 16(1)).

clause 37
FCEC clause 5(3)

- Clause 37 requires the Contractor to allow access at all times, for anyone authorised by the Engineer, to any place where work is being executed or from whence materials, manufactured articles and machinery are being obtained. This requirement is passed on to the Sub-contractor in clause 5(3) of the FCEC Form.

  The FCEC Form offers a general back-to-back arrangement, for the Contractor, vis-a-vis his obligations and responsibilities to the Employer, and those of the sub-contractor to him. However, this can only be true for any particular sub-contract if the Contractor vets the details of any sub-contract quote, and agrees any necessary changes with the Sub-contractor. For instance, domestic grouting sub-contract tenders may all be within a few per cent of each other, but individual rates may vary widely. If the sub-contractor's and Contractor's rates do not bear a constant relationship to each other, their respective final accounts for the work could be greatly different. A grouting sub-contractor may have priced to maximise the final account, by pricing injection of grout high, on the assumption that quantity for this will increase more than the quantity for drilling. Other tenderers may price the work completely straight. The Contractor needs to assess such differences before making his choice of sub-contractor.

Similar differences in rates may arise in other operations, particularly those for which accurate quantities cannot be obtained at the time of tender, but could also occur if there was a mistake in the original Bill of Quantities.

The Sub-contractor could also weight his rates and prices to obtain more of his profits and overheads on the first items to be carried out. Although the Contractor may follow the same practice, the Contractor's early work is not necessarily the same as that for the Sub-contractor.

Notes on the Sub-contract conditions

The FCEC Sub-contract Form consists of a number of standard clauses followed by five schedules which are intended to list the details for the particular sub-contract. These schedules are

(*a*) particulars of Main Contract
(*b*) details of the Sub-contract
(*c*) further particulars of the Sub-contract
(*d*) plant and facilities supplied by the Contractor
(*e*) insurances.

Comments on these schedules are given against the relevant clause numbers of the FCEC Form.

The FCEC Conditions, clauses

Clause 1 gives definitions and is self-explanatory.

Clause 2 is clear except for the matter of items supplied by the Contractor, which are dealt with under the comments on clause 4.

Clause 3 requires the Sub-contractor to perform the Sub-contract Works in such a way that will satisfy all the Contractor's obligations and responsibilities to the Employer, under the Main Contract. The Sub-contractor is deemed to have knowledge of all of the Main Contract, except the Contractor's prices. However, the Contractor needs to make sure that the Sub-contractor is given everything necessary to allow him to know exactly what is to be included in his price.

Clause 3(3) and 3(4) requires the sub-contractor to indemnify the Contractor against every liability he may incur owing to the Sub-contractor's breach of the Sub-contract. These liabilities include damages under the Main, or other contracts. The Sub-contractor's liability is for actual damage, but this would be nullified by the inclusion of Liquidated Damages in the Sub-contract.

The often-used argument is: if a Sub-contract is one tenth of the Main Contract in value, there should be a provision in the

Sub-contract for Liquidated Damages, at that fraction of those in the Main Contract. Where the delay of the Sub-contractor causes a delay to the completion of a section or the whole of the Works, the Contractor's costs would consist of: Liquidated Damages, if charged by the Employer, plus any justifiable costs for prolongation directly incurred by the Contractor or submitted by other sub-contractors. The Contractor would need to be very sure of a Sub-contractor's ability to perform in time, before he accepted Liquidated Damages into the Sub-contract, unless the figure was a genuine pre-estimate of the Contractor's likely damages.

Clause 4 deals with facilities the Contractor is to supply to the Sub-contractor. The use of standing scaffold is automatically included in the Sub-contract, unless clause 4(1) is omitted. The use of all other of the Contractor's construction plant and facilities is dependent on that use, and the terms of such use, being stated in the Fourth Schedule.

Part one of the schedule is for plant or facilities in common use, such as tower cranes and canteens, whereas part two is used to detail those items to be used exclusively by the Sub-contractor. The schedules must be filled in carefully to ensure that all the conditions that govern the supply and use of each item are included. In most cases there would appear to be little advantage in including exclusively used plant in the sub-contract, when it can be hired through a normal hire agreement.

There may be a need for the Contractor to supply some materials to the sub-contractor, but these do not fit into the Fourth Schedule, and should be included elsewhere in the documentation. If materials are to be supplied free of charge, some percentage for wastage should be incorporated.

The sub-contractor is required, under clause 4(4) to indemnify the Contractor against any damage or loss due to misuse of the plant or facilities supplied to the Contractor.

Clause 5 is concerned with the rules for working on Site, non-exclusive possession of those parts of the Site and access thereto, required for the Sub-contract Works. The Contractor, the Engineer and others authorised by them are given the right of access to the Sub-contract Works, workshops, working and storage areas on Site and places off-site, where materials or work are being prepared.

Clause 6(1) deals with commencement and completion, but can cause problems if the Contractor uses it to protect himself against his own difficulties in progressing in accordance with his programme.

The minimum 10 days' notice of the date of commencement, if insisted upon, is insufficient time to allow the Sub-contractor to organise his business, particularly if all contractors fixed starting dates in that way. By giving only 10 days notice, the Contractor may considerably reduce the Sub-contractor's opportunity for making a claim for a delayed start to the Sub-contract and by so doing, is increasing the problems of the Sub-contractor, who is likely to have less resources to cope with them.

The knowledge that he is protected against claims from sub-contractors might reduce the Site agent's commitment to maintain programme, in the early days of the Contract, when problems arise. This often results in exacerbation of problems of keeping to revised programmes later on.

In most cases, any design and/or manufactured items that the Sub-contractor has to produce before he starts work on Site will need to be completed before receipt of the 10 days' notice. This involves the Sub-contractor's being given a provisional date for commencement, in order to assess the viability of the timing for preparatory work.

If the necessary preparatory work before commencement on Site is particularly costly, and there are no arrangements for payment on account, the Sub-contractor may delay this work to minimise the cash outflow. This is particularly so where the Contractor has not been in touch with the Sub-contractor since placing the order. The Site agent should know all the lead-in timings and keep the Sub-contractor posted as to the latest start date for the commencement of preparatory work and work on Site. The principal objective of the exercise is to make money by making sure the work is done, not by making claims against the Sub-contractor.

The period for Completion for the Sub-contract works is entered in the Third Schedule, section C. In many cases there will need to be more than one period to tie in with any Section Completions in the Main Contract, or when a number of visits to the Site are necessary to complete work, in sequence, in conjunction with other operations. Where separate periods are necessary, the work to be completed in each period should be listed in one of the Sub-contract documents. Perhaps the best place to do this is in section (B) of the Second Schedule.

Clause 6(2). The causes of delay that entitle the Sub-contractor to an extension of time are given in clause 6(2). As would be expected these include those which would entitle the Contractor to an extension, under the Main Contract and variations to the Sub-contract (which do not give rise to an extension entitlement of the Main Contract) and breaches of the Sub-contract by the Contractor. The variations mentioned might include work which is not on the Main Contract critical path, or the removal of part of the Sub-contract work because of some error in information given by the Contractor, or additional work given to the Sub-contractor, which was in the original Contract. The most likely breach of the sub-contract, by the Contractor, is the failure to give possession of part of the Site to the Sub-contractor because preceding work has not been completed.

Where the Contractor would also have an entitlement to an extension, clause 6(2) requires, as a condition precedent to the Sub-contractor's right to a similar award, that he gives notice within 14 days of the cause of delay that would arise. This seems harsh, and would have no effect on the Sub-contractor, provided the Contractor obtained his extension, under the more generous notice provisions of the Main Contract. The Contractor would have no actual damages to claim under clause 3(3), and the Sub-contractor would be able to recover his additional time in the valuation of the variation (clause 9).

The final sub-clause of clause 6 requires the Contractor to notify the Sub-contractor, in writing, of all extensions to the Main Contract, which affect the Sub-contract. This is often done with great reluctance, but the Sub-contractor, as a part of the Contractor's team, should be kept informed of circumstances which require some adjustments to his performance, both on and off Site.

Clauses 7(1) and 7(2) describe two ways of issuing valid instructions or decisions under the Sub-contract

- those given by the Engineer or his representative and confirmed by the Contractor
- those given by the Contractor, who has the same powers to do so under the Sub-contract as the Engineer has under the Main Contract.

The point to note is that instructions are only valid if issued or confirmed by the Contractor, and he should make arrangements with the Engineer and the Sub-contractor to see that a workable system is operated, to avoid arguments. The Contractor should avoid involvement in any design by the Sub-contractor which is not within the Sub-contract.

The Sub-contractor is required to comply with all valid instructions and decisions and, in respect of these, has the same rights as the Contractor would have, if they had been given by the Engineer under the Main Contract. An exception applies to any invalid Engineer's instructions, confirmed by the Contractor, to the Sub-contractor. In such a situation, the Sub-contractor is entitled to any additional cost incurred, provided it was not caused or contributed to by any breach of the Sub-contract by the Sub-contractor.

Clause 8 covers similar ground to the previous clause, but with respect to variations. These must be given or confirmed, in writing, by the Contractor. As well as those instructions which constitute variations under the Main Contract, there are two other categories of variation under the Sub-contract

- changes to the Main Contract, agreed between Contractor and Employer, and confirmed to the Sub-contractor by the Contractor, e.g. the addition of a structure not contemplated in the original Contract or the acceptance of sub-standard work, under specified conditions
- direct instructions from the Contractor which could concern temporary works, or the addition of work to the Sub-contract which the Contractor originally intended to execute himself.

This clause specifically states that the Sub-contractor should take no action on instructions received by him, direct from the Engineer. The Contractor, after receiving a copy of such instructions from the Sub-contractor, should confirm the instruction and remind the Engineer of any previous agreement to restrict the issue of instructions to those persons, nominated by the Site agent to receive them.

This practice is more likely to occur with Nominated, rather than domestic Sub-contractors. The Contractor should not direct the Sub-contractor to refuse to accept instructions from the Engineer, but should at all times inform the Engineer that the Sub-contractor is unable to act on such instructions until confirmed by the Contractor.

Clause 9. In clause 9(1) the Sub-contract is stated to be a Lump Sum, with additions and omissions for variations, as against the Main Contract, which is totally remeasured. In practice the difference disappears, in many cases, because, where nothing to the contrary is stated in any bill of quantities, the difference between billed and actual quantities is measured and valued as an authorised variation (clause 9(4)). The value of all variations is derived from Bill rates, where possible, otherwise a value which is fair and reasonable in all the circumstances, which takes regard of any valuation for the same variation, comes under the Main Contract.

Most Contractors would contend that this entitlement is to have the varied rate, in the Sub-contract, calculated in the same manner as that for the Main Contract, so far as is possible, but the Sub-contractor has no right to know the Contractor's rates. The different allocations between preliminaries and rates, in the Contractor's and Sub-contractor's bills, can add complications to the use of the same method to derive rates.

There are a number of reasons why the measurement of a variation in the Sub- and Main contracts are different, including: a different method of measurement; the variation to the Main Contract is only partially carried out by the Sub-contract; the variation was known at the time of, and was incorporated in, the Sub-contract tender.

Sub-clause 9(5) contains the provisions for the execution of work on daywork, where the Contractor has authorised this method of payment to be used, for a variation. The alternatives given are the same as those included in the ICE *Conditions* i.e.

- pricing a schedule in the Bill of Quantities
- use of the Federation of Civil Engineering Contractors' Daywork Schedule.

The incorporation of the FCEC Schedule without adjustment will give the same payment for plant, already on Site, to both Contractor and Sub-contractor, as the Schedule rates apply to all plant on Site (see Amendment Sheet 1 to Schedules of Dayworks Carried out Incidental to Contract Work, edition dated August 1983).[4]

There may also be objections from the Engineer to paying the full schedule amount for labour plus $12\frac{1}{2}\%$ for a sub-contractor's invoice, and he may want all daywork done by the Contractor, to save the Employer the additional percentage.

The amount of supervision supplied by sub-contractors varies but it is difficult to justify entitlement to the same percentage on labour as the Contractor, so the incorporation of the FCEC Daywork Schedule, without adjustments, into sub-contracts, should be avoided. The better choice is the incorporation of a special schedule in the Sub-contract, even where this is not the case in the Main Contract.

There is no specific timing requirement for the submission of daywork documentation in clause 9, but the Sub-contractor is required to ensure that the Contractor can comply with the Main Contract timing, by clause 10.

Clause 10(1) requires the Sub-contractor to give all notices, accounts, returns or other information to the Contractor to enable the Contractor to comply punctually with the terms of the Main Contract, provided the requirement is known by the Sub-contractor. Where the timing and detail are stated in the Main Contract Documents the Sub-contractor is deemed to have knowledge of them, but this would not cover items such as 'Returns of Labour and Plant', which by clause 35 of the Main Contract need be submitted, to the detail requested, only if required by the Engineer. Any specific requirements such as this, known before sub-contract tender, should be incorporated into the Sub-contract Agreement, as are the dates for submission of interim statements. Otherwise they should be notified to the Sub-contractor as soon as possible. There is a specific time limit in the Sub-contract for the notification of delays that give rise to entitlement to an extension of time (notes on clause 6).

clause 35

Clause 10(2) imposes an obligation on the Contractor to secure, on behalf of the sub-contractor, any benefits claimable under the Main Contract, which apply to the Sub-contract, provided the Sub-contractor has given the proper notices timeously, and has supplied the necessary details. The benefits would be those claimable by the Contractor under clause 52(4)(b) of the ICE *Conditions*. Where such benefits are received by the Contractor, he is required to pass on such proportion as is reasonable to the Sub-contractor.

clause 52(4)

Clause 10(3) allows the Contractor to make deductions from payments to the Sub-contractor, if non-compliance with clause 10(1) prevents recovery by the Contractor of any sum from the Employer. In this event the Sub-contractor would be entitled to written details of the default, and a detailed account of the reasons for, and amount of, the deduction.

Clause 11 contains provisions re the property in materials and plant which allows the Contractor to comply with clause 53 of the Main Contract.

clause 53

Clause 12 requires the indemnities that are given to the Contractor, by the Sub-contractor, and vice versa, to be in line with those between Employer and Contractor, under clauses 22 and 24 of the ICE *Conditions*.

clause 22
clause 24

Clause 13. The Sub-contractor is required to maintain the Sub-contract Works in the condition required by the Main

Contract, until the relevant section or overall completion of the Main Contract Works. Then, he is subject to the same maintenance period and requirements as those for the Main Works. It seems hard that the Sub-contractor, in the interim between his completion and the relevant Main Contract completion, is responsible for all damage except fair wear and tear and that caused by the Contractor, the Employer or his servants or agents. The Sub-contractor will not be on Site and, unless responsibility is accepted voluntarily by someone, he may find it difficult to prove who caused the damage.

Clause 14. The insurance cover, required of the Sub-contractor, is not particularised in the clause, but is left to be fixed, in each case, by inserting the requirements in parts 1 and 2 of the Fifth Schedule.

The Contractor gives details of the Insurances required of the Sub-contractor in part 1, which should include 'insurance of the Sub-contract Works' unless this is shown to be covered by the Contractor's policy by the details of that policy being given in part 2. Contractors usually require Sub-contractors to insure the Sub-contract Works.

If it is intended that the Sub-contractor should have the benefit of the Main Contractor's 'Insurance of the Works' policy the wording entered into part 2 should avoid reference to clause 21 of the ICE *Conditions*, since that encompasses several insurances e.g. the Works, plant, motor vehicles and professional indemnity.

clause 21

In stating, in part 1 of the Fifth Schedule, the insurances to be taken out by the Sub-contractor, the Contractor should take into account the particular circumstances. Apart from Works All Risks and Public Liability Insurance for the same minimum as is stated in the Main Contract Form of Tender (Appendix), professional indemnity insurance to cover design of part of the Temporary or Permanent Works, or insurance of materials stored off the Site, may be required. Employer's Liability Insurance is a statutory requirement, and need not be listed.

Form of Tender (Appendix)

All insurances should be inspected to confirm that they comply with the requirements, and proof that the premium has been paid should be requested. It is important, at initial inspection, to note premium renewal dates and check that the insurance has been renewed. This inspection of policies and premium payment can be extremely onerous, because, in the absence of standard wording, the policies are complicated and difficult to understand by people other than experts in the insurance field.

Clause 15 is by far the longest clause in the sub-contract conditions, because it is complete in itself, rather than a reference to the Main Contract conditions. The timing of the Sub-contractor's interim statements and payments are referenced to the 'Specified Dates' which are the dates on which the Contractor intends to submit his interim statements under clause 60 of the Main (ICE) Contract Conditions.

clause 60

The First Schedule contains a section for entering the 'Specified Dates', and if these were not available when the Sub-contract was formed, they should be given to the Sub-contractor as soon as the end of measurement periods are agreed with the Engineer. The Sub-contractor should also be given the form of interim statements and the details required in them, which will probably be based on those required for the Main Contract (clause 60 of the Main Contract conditions, and clause 15(1) of the FCEC Form). The Sub-contractor's interim statement is referred to as a 'Valid Statement'.

clause 60

Clause 15(1) (a) only mentions the inclusion of the value of work done and qualifying materials in the Valid Statement, but the Sub-contractor could be entitled to payments other than for these items, such as

- costs for disruption and delay, which can be included in the Main Contractor's statements, under clause 60(1) (d) and to which the Sub-contractor can be entitled under clause 10(2) of the FCEC Form
- cost of any non-valid instruction of the Engineer or his Representative, confirmed by the Contractor, if the proviso in clause 7(1) is met
- cost of any instruction given by the Contractor, if a corresponding instruction from the Engineer to the Contractor would have entitled the Contractor to recover the cost (clause 7(2))
- damages for any breach of the Sub-contract by the Contractor.

clause 60(1)

It may be that the Contractor wishes to keep the last three items in a separate account as they would not be directly recoverable from the Employer, but the Sub-contractor would need some indication of requirements, and how and when he would be paid.

The Sub-contractor submits his statement seven days before, and is paid 35 days after, the Specified Date. If both the Employer and the Contractor pay on the last allowable day, the interval between the two payments is seven days. The 35-day period will require adjustment if the 28-day period in clause 60 is amended.

clause 60

Sub-contract payments are subject to the retention percentages and limit inserted in the Third Schedule. It is impossible to insert a general limit throughout the Contract that avoids any discrepancy between the Contractor's and Sub-contractor's proportion of retention. The position as regards retention is complicated because the percentage retained on materials is called retention in the Sub-contract, but not in the Contract. The percentage retention for materials, which is usually higher than that for work, would result in the limit of retention's being reached quickly on a Sub-contract, with a substantial material element.

A fair compromise is to make the retention limit for the Sub-contract equal to the retention on materials plus either 3% of the Sub-contract Sum or £1500, whichever was the greater. This would be the retention on a Main Contract of the same value as the Sub-contract.

The amount due for payment to the Sub-contractor is subject to deductions, which the Contractor is authorised to make by clauses 3(4), 10(3) and 17(3) of the FCEC Form, following a default on the part of the Sub-contractor. The Contractor should give the Sub-contractor full written details of the reasons for, and amounts of, any deductions made.

Form of Tender (Appendix)

No interim payment is due to the Sub-contractor if the value of the Contractor's application is less than the 'Minimum Amount of Interim Certificate' on the Form of Tender (Appendix) in the Main Contract or if the Sub-contract interim value is less than the minimum in the Sub-contract. There is no prepared spot in any of the five schedules to insert a minimum amount for a Sub-contract payment, and there seems no reason for such a limitation. If, to ensure a back-to-back, the minimum interim value for the Sub-contract is written in 'As main Contract' this, in many cases, would be more than the value of the Sub-contract, and quite unfair to apply, if unnoticed by the Sub-contractor.

The Sub-contractor is not entitled to payment in respect of any item not certified by the Engineer, or certified and not paid by the Employer, or where a dispute has arisen between the Sub-contractor and the Contractor and/or the Contractor and the Employer on any matter included in the Sub-contractor's statement. The Contractor should not use these reasons for non-payment unfairly, so as to cause a cash flow problem. For example, if both Contractor and Sub-contractor are carrying out earthworks, any reduction in quantity, made by the Engineer in the quantity in his interim certificate, should be allocated properly between the Contractor and Sub-contractor, not assumed to be all against the Sub-contractor's work.

Nominated Sub-contractors are given some protection from unwarranted reductions in the amount certified for them, paid by the Contractor, by the provisions of clause 59C of the ICE *Conditions*. This clause gives the Employer the right to make direct payments to a Nominated Sub-contractor, against the Engineer's certificate, on all amounts underpaid in previous certificates. The Engineer may only issue a direct payment certificate if the Contractor has failed to supply reasonable proof that payment has been made or, failing that, has supplied both Engineer and Sub-contractor with a reasonable cause for non-payment. The same procedure must be used before each interim certificate, to establish any further entitlement to direct payment.

clause 59C

The decision on whether or not to make a direct payment usually hinges on the Contractor's given reason for non-payment. The procedure is rarely used because of the danger of the Employer's ending up by paying both Contractor and Sub-contractor, for the same work, if things go wrong. However, the Contractor may lose the sympathy of the Engineer, if he suspects that the Contractor is acting unfairly.

The right of the Sub-contractor to receive interest on overdue payments from the Contractor is not automatic, as in the main Contract, but is subject to the Sub-contractor making a claim, in writing (clause 15(3) (f)). However, the Sub-contractor is entitled to any interest paid to the Contractor, in respect of late payment for Sub-contract work, even though a claim was not submitted (clause 15(3) (g)).

Release of the first half of retention to the Sub-contractor is dependent on completion of the relevant section of the Main Works, not the completion of the Sub-contract Works.

Clause 15(5) deals with the payment of the Sub-contractor's statement of final account, which cannot be paid in full until the release of the second half of retention, following completion of the maintenance obligations. However, the actual final account for many sub-contractors can be completed during the construction period or soon after, and Sub-contractors should be encouraged to do so.

The Sub-contractor is required by clause 15(6) to make any claims before the issue of the Maintenance Certificate for the last Section which involves part of the Sub-contract Works and not the Maintenance Certificate for the Whole of the Works. This is unhelpful, because the Engineer is required to issue one Maintenance Certificate only, and that is when the whole of the Works have been completed and maintained (clause 61). If clause 15(6) is to have a meaning, the reference in it to 'the

clause 61

Maintenance Certificate in respect of the last of such Sections' should be amended to 'the end of the Maintenance Period for the last of such Sections'. This would give the Sub-contractor some extra incentive to complete his final account, as only by doing so would he be sure that all extras had been considered.

Clause 16(1) and 16(2) cover determination of the Main Contract, not caused by the Sub-contractor, and gives the Sub-contractor the same rights the Contractor would have if the Main Contract was determined, under clauses 64 or 65 of the ICE *Conditions*. The Sub-contract is not automatically determined on the determination of the Main Contract.

clause 64
clause 65

Clause 16(3) deals with the situation where the Sub-contractor's default is the cause of determination of the Main Contract. In such a case the Sub-contractor will be liable for any loss or cost the Contractor will incur because of the action taken by the Employer, under clause 63.

clause 63

Clause 17. The list of defaults which allow the Contractor to determine the Sub-contractor's employment, under the Sub-contract, are similar, but not identical, to those listed in clause 63 of the ICE *Conditions*. Abandonment of the Sub-contract is not included in clause 17, but as has been mentioned in the comments on clause 63 (chapter 4) this is unlikely to cause a problem.

clause 63

There are no time limits in clause 17(1) (a)–(c) within which the Sub-contractor has to comply with the Contractor's notice to remedy a default, which means that the Sub-contractor would have a reasonable time to do so.

Many of the problems which arise in determination of sub-contracts come from the incorrect application of the procedure for so doing, by a Site agent, who is over-anxious for better progress on the Works. If the Contractor gets the procedure right, he still has to decide whether or not the Sub-contractor has complied with the prior notice. This is easy enough if the notice is with regard to work or materials, which both parties accept do not comply with the Contract, but there is still the matter of time allowed for compliance. From a practical point of view 'reasonable time' would depend on the effect of non-compliance on the continuity of working on other operations and whether a similar notice had been given under the Main Contract.

If the notice, under clause 17(1) (a), was with regard to failure to proceed with due diligence, and the Contractor was of the opinion that the Sub-contractor had taken all possible steps to improve his performance, there would be no advantage in determining the sub-contract, even though completion would not be achieved in time. If a Nominated Sub-contract was in this position, the Contractor would seek an instruction from the Engineer (see under Nominated Sub-contractors).

The Contractor has an alternative to determining the Sub-contract, when he is entitled to do so. Under clause 1/(3) he is allowed to take part of the Sub-contract Works from the Sub-contractor, and either execute this himself or employ someone else to do so. Before taking this action the Sub-contractor must have been given written notice to remedy a default, and have failed to do as required. As with determination, the only reason for taking part of the Sub-contract Works away from the Sub-contractor, when entitled to do so, is to improve the overall situation on the Contract.

The Site agent should monitor Sub-contractor's progress to ensure that corrective measures, if necessary, can be taken early enough to have the greatest chance of success. In this respect, Sub-contractors, who are following a latest start line on a critical path programme, should be asked to indicate the resources required to attain a successful completion, within the required time. Latest start dates are associated with increasing resource levels, as time passes, with the consequent greater difficulty of maintaining the correct rate of progress (see chapter 3).

Clause 18. The disputes clause, 18, requires disputes to be settled by arbitration, between Contractor and Sub-contractor, but allows the Contractor to require the Sub-contractor to join in an arbitration, on the same points, under the main Contractor, provided that no notice of arbitration has previously been given under the Sub-contract. The Site agent is in the position to gauge the Sub-contractor's intentions with regard to taking arbitration, under this clause, and can obtain advice before any action is taken by the Sub-contractor.

Clause 19. The Sub-contractor is in the same position with regard to VAT as is the Contractor, in that he does not allow VAT in his prices, and can recover any positive tax chargeable on his taxable supplies.

Appendix 1. Notices to be given by the Contractor

Notice	Relevant clause	Timing
Reference to the Engineer (E) in writing of an Engineer's Representative's (ER's) decision for confirmation or amendment	2(4)	No timing given in clause 2 but argument will be avoided if time limits for contesting an Engineer's decision on the same subject are maintained
Request in writing to the E for permission to sub-contract part of the Works	4	Allowing time to give consent before it is necessary to sub-let
Request the E or the ER to resolve a discrepancy or ambiguity in the contract documents	5 13(3) 44(1) 52(4)(b)	Whenever they come to light
Request the E for more than two copies of the drawings, if needed	6	As required
Request the E or ER in writing to supply further drawings required for the execution of the Works	7(2)	So as to give adequate notice of the requirement
Notice in writing to the E or ER that delay in the issue of drawings is causing delay on additional cost	7(3) 44(1) 52(4)(b)	Within 28 days of the delay, or as soon as is reasonably possible
Notice in writing to the E that unanticipated physical conditions or artificial obstructions are causing delay or additional cost	12(1) 44(1) 52(4)(b)	As soon as reasonably possible
Notice that the E's or ER's instructions given under clause 13(1) are causing delay or additional cost	13(1) 13(3) 44(1) 42(4)(b)	As soon as reasonably possible following the instruction
Submit his programme for carrying out the Works to the E or ER	14(1)	Within 21 days of the acceptance of his tender

Provide the E or ER with a general description of the arrangements and methods of construction to be used	14(1)	Within 21 days of the acceptance of his tender
Furnish such details as the E or ER may request concerning the programme	14(1)	Within a reasonable time after the request
Submit a revised programme to the E or ER modified to show how the Works are to be completed in the time or extended time for completion	14(2)	On request of the E or ER when progress does not match the approved programme
Submit to the E or ER details of the methods of construction, including Temporary Works, and calculations of the stresses and strains that will be set up in the Permanent Works	14(3)	On the request of the E or ER and at such times as they may reasonably require
Request the E or ER's consent to changes to the methods of construction or Temporary Works required by the E or ER	14(4)	Within a reasonable time of being informed how the Contractor's original proposals need to be amended to meet the E or ER's requirements
If the E's approval of the methods of construction is unreasonably delayed or limitations imposed cause delay and/or extra cost give notification to that effect, and make a claim for extra cost and extension of time	14(6) 44(1) 52(4)	Submission to the E for extension, within 28 days or as soon as reasonably possible after. Submission to the E or ER for additional cost as soon as possible after events giving rise to the claim have occurred
Request for approval in writing to the Contractor's agent for the site	15(1)	As soon as the name of the person to be appointed is known
Request for additional data required for setting out to be given in writing	17	As the need arises to the E or ER
Report any damage to the Works caused by the Excepted Risks	20(2)	As soon as possible after such damage has occurred, and to the E or ER
Request approval for the proposed insurers for the Works and the terms of the policy	21	Before insuring the Works

Submit the insurance for the Works and receipt for the current premium	21	Whenever required by the Employer
Give details of any damage to third party property caused by the provisos in clause 22	22(1) 22(2)	To the Employer when such damage occurs, with a copy to the E or ER
Request approval of the proposed third party insurance policy and insurer	23(2)	Whenever required by the Employer; normally the E or ER would vet the policy on behalf of the Employer
Produce the third party insurance policy and receipt for the current premium	23(2)	Whenever required by the Employer
Give details of any accident or injury to workmen caused by the default of the Employer, his agents or servants	24	To the Employer as such injuries or accidents occur; the E or ER should have a copy as he will probably be involved
Give details of any fees paid and request inclusion in the next certificate	26(1)	To the E or ER following payment of these by the Contractor
Give notice of proposed commencement of works in a street or controlled land	27(4) 27(5)	To the Employer at least 21 days before the proposed commencement; (in practice the Engineer may well be authorised to deal with this on behalf of the Employer)
Submit notice of any delay or extra cost in streets or in controlled land	27(6) 44(1)	To the E within the time limits in clause 52(4)(a) caused by variations, 44(1) for extensions and 52(4)(b) for additional costs
Notify the E of any claims received for damage to highways or bridges	30(3)	To the E or ER as soon as such claims are received by the Contractor
Submit notice of any delay or extra costs caused by the Employer's other Contractors	30(2) 44(1) 52(4)	To the E or ER within the time limits in clause 44(1) for extensions and 52(4)(b) for delays causing extra costs
Give notice of the discovery of any fossils, coins or things of geological or archaeological interest	32	To the E or ER immediately on discovery and before removal
Submit labour and plant returns	35	To the E or ER in the form and at the intervals he prescribes

Give notice that work is available for inspection	38(1)	To the E or ER when the work is ready and before it is covered up
Give notice rejecting allegations of faulty materials or workmanship, if these are considered to comply with the Contract	39(1)	To the E or ER before removal of the alleged faulty material or workmanship
Give notice if the issue of a suspension order is the cause of delay leading to additional costs or the need for an extension of time	40(1) 44(1) 52(4)(b)	To the E or ER within the time limits in clause 44(1) for extension and to 52(4)(b) for additional costs
Give notice requesting permission to re-start suspended operations	40(2)	To the E after a suspension order has been in force for at least three months
At his discretion, give notice that the Contractor will take the failure to remove a suspension order as an omission of the affected part of the Works from the Contract	40(2)	To the E, at least 28 days after the notice requesting permission to re-start was received by the Engineer
Give notice that non-possession of part of the Site is causing delay and/or additional costs	42(1)	To the E or ER within the time limits in clause 44(1) for delays leading to an extension of time and 52(4)(b) for extra costs
Submit details of any extension of time entitlement	44(1)	To the E, within 28 days of the cause of delay having arisen, or as soon after as is reasonable
Request permission to work at night or weekends, or, if such working is necessary in an emergency give notification that work is to be executed	45	To the E or ER before the work being executed, or in an emergency immediately the need is known
Submit details of the steps being taken to expedite progress	46	To the E or ER after being notified that progress is too slow to ensure completion in accordance with the Contract
Request a completion certificate for the whole, a part or a section of the Works, giving an undertaking to complete any outstanding work during the Maintenance Period	48	To the E when, in the Contractor's opinion, substantial completion has been achieved

Give notice that a variation makes other rates or prices unreasonable or inapplicable	52(2)	To the E or ER before the work is commenced or as soon thereafter as is reasonable
Submit quotations for materials to be used on daywork	52(3)	To the E or ER before ordering the materials
Submit a duplicate list of all operators' names, occupations and times spent on daywork and a statement giving a description and quantities of all materials used	52(3)	To the E or ER daily while that work is being executed
Submit receipts or other proof of amounts paid out for daywork operations or materials	52(3)	To the E or ER presumably with the priced daywork schedule at the end of each month
Submit a priced out schedule of work done on daywork	52(3)	To the E or ER at the end of each month in which dayworks are executed
Give notice of intention to claim a higher rate or price than that fixed under clause 52(1) or (2) or clause 56(2)	52(4)(a)	To the E or ER within 28 days of the notification that the rate or price has been fixed
Submit notice of intent to claim additional payment under any clause except clauses (52(1) or (2)	52(4)(b)	To the E or ER as soon as reasonably possible after the happening of the event giving rise to the claim
Supply copies of records maintained in support of claims	52(4)(c)	To the E or ER as and when required
Submit full details of the grounds for claims and the amount claimed	52(4)(d)	To the E or ER as soon as reasonable and thereafter at such intervals as the E or ER may reasonably require
Notify the name and address of the owner of any item of plant and certify that the agreement for hire allows the Employer to take over the hire, in the event of a forfeiture, under clause 63	53(5)	To the E or ER on request
Request permission to remove plant from the Site	53(6)	To the E or ER before removing the item of plant from the site

Provide documentary evidence that the property in goods or materials, stored off-site, for which payment is allowed, is vested in the Contractor and a detailed list of the description, value and location of these items	54(2)	To the E or ER in time to allow payment for the materials in the next certificate
Furnish particulars in relation to any work being measured	56(3)	To the E or ER as and when required
Produce all relevant details and documentation with regard to expenditure on work by Nominated Subcontractors	58(6)	To the E or ER when required
Give notice of objection to the nomination of a subcontractor	59(1)	To the E or ER as soon as reasonably possible after nomination and before placing any order
Give notice that events have occurred making a subcontractor liable to forfeiture of the subcontract	59(B) (2)	To the E or ER before taking any action with regard to the forfeiture
Supply proof of payment to a sub-contractor or give reasons for withholding payment	59(C)	To the E or ER on demand
Submit a statement showing details of the amount of payment thought to be due	60(1)	To the E or ER after the end of each month
Submit a statement giving full details of the final account	60(3)	To the E not later than three months after the date of the Maintenance Certificate
In the event of Contract forfeiture, assign any sub-contracts or goods or materials supply agreements to the Employer	63(2)	To the Employer by way of the E, forthwith on being given notice to do so
Request the E's decision on a dispute	66(1)	To the E but no time limit is given
If dissatisfied with the Engineer's decision or if no decision has been given in the time limit, give notice of referral of the dispute to an arbitrator for settlement	66(3)	To the Employer and the E within three calendar months of receipt of the decision, or the end of the period in which such decision should have been given

Give notice of a change in the level or incidence of a labour-tax matter, if clause 69 applies to the Contract	69(6)	To the E or ER as soon as is practicable after the event affecting the costs has occurred
Issue a VAT tax receipt in respect of any payment for VAT included in interim or final payments	70	To the Employer within seven days of any payment including VAT

Appendix 2. Check list for the Contractor/agent

Clause in ICE *Conditions*	Action required
	As soon as acceptance has been received check that details contained in the acceptance are in line with the Tender and any subsequent amendments. Take necessary steps to correct discrepancies found.
2(2)	On acceptance write to the Engineer asking for a meeting to discuss general matters in connection with setting up the various levels of communication and responsibility in the Contractor's and Engineer's Site and other office organisations. The proposed Agent should be introduced at the meeting.
15(2)	Following the meeting the Contractor should write formally, to the Engineer, giving name, qualifications and experience of the proposed Site Agent, and request approval, in writing, for the appointment. As appointments are made, the Engineer and his Representative should be kept informed of the qualifications of the senior staff. On appointment the Agent should arrange a meeting with the estimating department to go through the details of the tender pricing, methods of working and programme on which the tender was based. Ask that the breakdown of the Tender be adjusted for any last minute adjustments to the Tender. If the appointment is the first as an Agent the Agent should ask his superior to introduce him to the head office service people, responsible for services to the Site, e.g., plant, safety, accounts, industrial relations, measurement, computer services and temporary works design.
2(3)	If no confirmation of the authority delegated to the Engineer's Representatives has been received, write asking for such authority, so that an appropriate site meeting can be arranged to discuss the respective Site organisations.
14(1)	Within 21 days after acceptance of the Tender, submit the programme for carrying out the Works.

41	Check that the Engineer's letter confirming the Date for Commencement of the Works has been received, and is considered to be within a reasonable time of acceptance. If the Engineer's Representative has not arranged a first site meeting to discuss the site organisations, telephone to make arrangements, and confirm the date, giving a list of the items that you require on the agenda for the meeting. Fix meetings with British Telecom, British Gas, Electricity, Water Authorities, and place orders for Site services.
	Points to establish at the meeting or soon thereafter
2(1), 2(2)	Contractor's site staffing arrangements. Engineer's Representative's site staffing arrangements. Offices, laboratories, workshops, Methods of Working and programme. Assistance required for the Engineer's Representative.
38(1)	Arrangements for obtaining approval of work, before it is covered up.
4	Work which it is proposed should be sub-contracted.
31	Confirmation that the only Employer's workmen or contractors to be on Site are those which are indicated in the Contract documents.
35	The form of labour and plant return, if any, that is required.
60(1)	Form of interim statements, if none is prescribed in the Specification.
60(1)	Agreement of month end dates for interim statements; ensure that the dates of any committee meetings, at which interim certificates are passed, are known. Agreement of the type, intervals between, items for discussion and agreement, at regular site meetings and also those who will attend for certain subjects. Progress Meetings, Rate Fixing Meetings and Meetings with other Authorities.
23, 23	Ensure that the Employer's approval is obtained for the insurer and the terms of the Contractor's insurance of the Works and against third party claims.
10	Ensure that the Bond, if any is required, is being satisfactorily progressed and is supplied as soon as possible.
42(2)	Make arrangements for the purchase or rent of additional wayleaves or land required for the Works. Draw up a schedule of numbers and categories of staff and labour required throughout the Contract, and how these requirements are to be met. Organise records to be kept, and ensure that the quality of those records are monitored from time to time. Arrange to visit local union representatives, but only in accordance with company policy, and advice. Set up a site incentive bonus scheme.

Draw up a schedule of the types and numbers of plant required throughout the Contact, and the source from which it is intended to hire them.

Draw up a schedule of work to be sub-contracted and obtain the Engineer's written consent for sub-letting.

Arrange to have sub-contract orders placed. Give sub-contractors details of Specified Dates; form of, and details to be contained in, Valid Statements; details of labour and plant returns required.

Monitor off-site manufacture and fabrication to ensure that on-site construction can start to programme.

Draw up a schedule of materials and their suppliers, with dates and quantities to be delivered, and place the necessary orders. Monitor progress of manufacture and fabrication to ensure materials will be available when required.

Organise responsibilities for Site safety, and those who are to keep the necessary records, etc.

Make liaison visits to organisations such as police, traffic wardens and local papers to open up avenues for liaison on future problems that may arise.

Appendix 3. Clauses relevant to extension of time

Events giving rise to extension entitlement

Clause	Event
7(3)	Delay in issue by the Engineer of any instructions or drawings requested by the Contractor, by adequate notice in writing
12(1), 2 (3)	Delay caused by physical conditions or artificial obstructions, notified by the Contractor, to the Engineer, in writing
5, 13(3)	Delay or disruption to the Contractor's arrangements or construction methods, caused by the Engineer's instructions, including those clarifying an ambiguity in, or discrepancies between Contract documents, provided such a delay or disruption is beyond that reasonably foreseeable by an experienced Contractor, at the time of tender
14(6)	Unreasonable delay, on the part of the Engineer, in approving the Contractor's proposed construction methods, or delay caused through a limitation to those methods, imposed by the Engineer, and which could not have been reasonably foreseen by an experienced contractor at the time of tender
27(6)	Any delay to the Works caused by the necessity to give notice, before carrying out work against a variation order, in a highway
31(2)	Delays caused by the Employer's own workmen or other contractors, and which are greater than those which could have been foreseen by an experienced contractor at the time of tender
40(1)	Delay caused by a suspension order issued by the Engineer, provided that it is not necessary for the execution or safety of the Works or because of weather conditions
42(1)	Delay to the Works caused by the Employer's delay in giving the Contractor possession of parts of the Site to allow work to be executed in accordance with the contract programme
44(1)	Delays caused by Variations ordered under clause 51(1) Increased quantities referred to in clause 51(3) and 56(2) Exceptionally adverse weather conditions Other special circumstances of any kind whatsoever

59B(4)	Notice of forfeiture on a Nominated Sub-contractor which is given in circumstances entitling the Contractor to do so, or is given with the Employer's consent, or at the direction of the Engineer

Procedure for claims and awards of extensions of time

Clause	Procedure
44(1)	Within 28 days of the occurrence of any of the events listed, or as soon thereafter as is reasonable, the Contractor shall submit his fully detailed claim to extension of time to the Engineer
44(2)	After consideration and assessment of the claim, the Engineer shall either make an award of an extension in writing or inform the Contractor that he has no entitlement. The Engineer can also award an extension without a claim from the Contractor
44(3)	The Engineer should review the situation as soon after the passing of any current completion date for a section or the whole of the Works, award appropriate extensions or inform the Contractor of a nil entitlement. The Engineer must take this action, with or without a claim from the Contractor
44(4)	The Engineer should make a final review of all circumstances on issue of any Certificate of Completion, for a section or the whole of the Works and make his final award, if any. He may not reduce extensions already given in such an award

Appendix 4. Calculation of Liquidated Damages

The amounts of Liquidated Damages payable, at various stages of completion for a Contract, which includes a number of sections, is easier to understand through examples. The inconsistencies involved in the use of the formulae included in clause 47 in arriving at genuine pre-estimated damages, are also best illustrated through examples. This Appendix contains illustrations of the above two points.

Take the example of a contract with three sections, in addition to the remainder of the work, which is completed on the date for the Whole of the Works. The various figures inserted into the Form of Tender (Appendix) would be represented by

Liquidated Damages for the Whole of the Works column (col) 1
Reduction to damages if sections are completed at the date for completion of the Whole of the Works

Section 1	col 2/1
Section 2	col 2/2
Section 3	col 2/3

Liquidated Damages for each section

Section 1	col 3/1
Section 2	col 3/2
Section 3	col 3/3

Then on the basis for calculating damages set out in clause 47(2)

(*a*) If any section is incomplete after the relevant date or extended date for its completion, damages are chargeable at the daily or weekly rate for that section, for the number of days/weeks that it remains incompleted.
If the completion dates for all three sections have passed, without any section having been completed, but within the Time for Completion for the whole of the Works, the total Liquidated Damages chargeable per day/week, while that situation exists is

col 3/1 + col 3/2 + col 3/3

When any section is completed, the relevant column 3 figure is removed from the calculation.

(*b*) Liquidated Damages chargeable per day/week if all sections remain incomplete after the date for completion of the whole of the Works has passed

col 1 + col 3/1 – col 2/1 + col 3/2 – col 2/2 +
col 3/3 – col 2/3

At the completion of any Section the relevant column 3 figure is removed from the formula, but the column 2 amount remains.

(*c*) Liquidated Damages chargeable, per day/week, if all sections are completed, but the whole of the Works remain incomplete, when the due date for completion of the whole of the Works has passed

col 1 – col 2/1 – col 2/2 – col 2/3

From (*c*), it can be seen that if none of the Works can be productive until the whole has been completed, then col 2/1, col 2/2 and col 2/3 should all be zero, leaving the damages chargeable the same, when the date for completion of the Whole of the Works has passed, whether or not any sections have been completed.

From (*b*), it will be seen that: if the values in col 3 are smaller than those in col 2, the calculation becomes a nonsense, since the total damages will be less than the rate in col 1, even though none of the Works have been completed.

One tender seen in the early days of the ICE *Conditions* 5th Edition had 13 sections all with zero in the relevant col 3, and £3000 in col 2, while damages for the whole of the Works, in col 1, was also £3000. The Contract had a construction period of only 19 months, which was extremely tight. However, the Contractor had nothing to fear from the application of the Liquidated Damages formula. On the contrary, he could look forward to the prospect of receiving £36 000 for every week, after the end of the Time for Completion of the whole of the Works, providing no sections were complete. Thereafter, the Contractor's windfall would be reduced by £3000, every time he completed a Section. It is not known whether the successful Contractor had these tender Liquidated Damages incorporated into the actual Contract, but, even if he had, he would have had difficulty collecting any money.

If there are particular damages for a section, these should be the difference between cols 3 and 2 for that section. If col 2 is zero then col 3 will be the special damages, such as the continuing supervision of that section plus the claims for delay of the Employer's other contractors, depending on the completion of that section.

If a part of a section is certified as complete, before the completion of the whole of the section, then in any calculation of damages due, col 3 and col 2 rates are reduced by the fraction of the original amount, represented by the value of the part, divided by the value of the section, and col 1 is reduced by the same amount as col 2 (clause 4/(2) (b) (ii)).

It is of obvious advantage to the Contractor to obtain as many partial completion certificates as possible, but he is only likely to obtain these if the part can be used without the remainder of the section or the whole of the Works.

Despite the complexity of the ICE Liquidated Damages clause, the

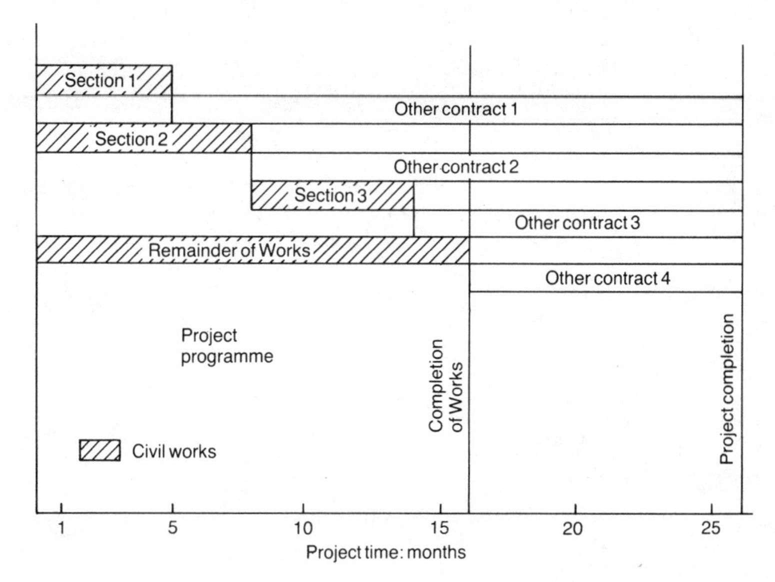

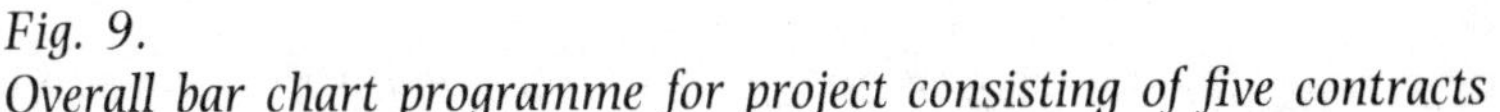

Fig. 9.
Overall bar chart programme for project consisting of five contracts

various formulae do not give a genuine pre-estimate of the damages likely to be incurred in all situations, as a few examples will demonstrate.

The simplest case, which the various formulae fit quite well, is of a contract, which is a complete project in its own right, and where parts of it can be used on their own. An example is a new road bypass on a green field site. If side road crossings were not included in the contract as sections, then partial completion certificates could be applied for each of these, and the same would also apply to each of the carriageways.

If the overall damages are based on the cost of the contract, plus land, plus supervision costs, the pro-ratas to calculate the reduction in damages would approximate to the correct figure.

In the example of a project programme, given in Fig. 9; a number of separate contracts make up one project. The consequnces to the Employer of delay in sections, or the whole of the civil works, would differ if each of the other contracts led to separate production units, or they all contributed to only one such unit. In either case, an equal delay to any section would delay the completion of the overall project by the same amount, and the overall project could be delayed, without a delay to the completion of the whole of the civil works.

If production on the project could start only when all contracts were completed, a delay to any Section, or the whole of the Works, would cause lost production for the period of the delay. The cost per day or week of the loss, and the cost to the Employer of the Engineer's supervision, and delays to other contractors, would need to be included in the calculation of damages, to be inserted in the civil Contract.

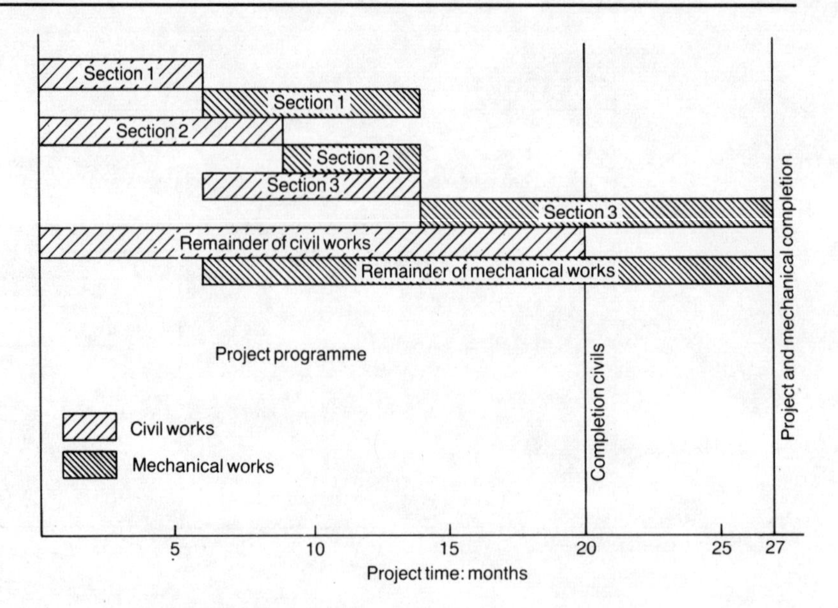

Fig. 10. Overall programme of civil and mechanical works for a project

If the following costs applied

cost of lost production	20 units/day
delay to Contract 1	8 units/day
delay to Contract 2	6 units/day
delay to Contract 3	5 units/day
delay to Contract 4	3 units/day

In the event of delay in the civil works to any one section, the damages deducted should be 20 units plus the relevant other contract delay cost. Therefore col 3 of the Form of Tender (Appendix) would be

Form of Tender (Appendix)

for Section 1	28 units/day
for Section 2	26 units/day
for Section 3	25 units/day
for the whole of the works	23 units/day

However, if both Sections 1 and 2 were in default simultaneously, the damages deductible, in accordance with clause 4(2)(a), would be $28 + 26 = 54$ units/day. This is 20 units too much, because the cost of the loss of production has been included twice.

If the 20 units/day is allocated for col 2 for each Section, then once the Time for Completion for the whole of the Works, has passed, the Liquidated Damages formula gives a reasonable answer, only if all sections remain incomplete.

If in the example shown in Fig. 9 a production unit became available at the end of each of the other contracts, the Liquidated Damages formula works for all combinations of completion of

Form of Tender (Appendix)

sections and the whole of the Works. Figures in the Form of Tender (Appendix) for the same values of work as before, could be

Damages for the whole of the Works 24 units/day

	col 2 units/day	col 3 units/day
Section 1	5	13
Section 2	6	12
Section 3	4	9

Figure 10 shows a project programme where the completion of the sections of the civil work control completion of the mechanical work and therefore the project as a whole. Some delay to overall completion of the civil work can be tolerated as long as the sections are completed in time. Clause 47 does not allow the calculation of Liquidated Damages to represent a genuine pre-estimate of the damages that are likely to result from delays to Sections and the whole of the Works.

References

1. *Conditions of contract and forms of tender, agreement and bond for use in connection with works of civil engineering construction.* Institution of Civil Engineers, London, 1973 (revised 1979), 5th edn.
2. Institution of Civil Engineers/Association of Consulting Engineers/Federation of Civil Engineering Contractors. *Guidance on the preparation, submission and consideration of tenders for civil engineering contracts recommended for use in the United Kingdom.* Institution of Civil Engineers, London, chapters 1 and 5.
3. *Civil engineering standard method of measurement.* Institution of Civil Engineers, chapter 2, pp. 36, 39 and 40; chapter 5.
4. *Schedules of dayworks carried out incidental to contract work.* Federation of Civil Engineering Contractors, London, 1983 (amended 1987), chapter 5.
5. Department of Transport. *Method of measurement for roads and bridges.* HMSO, London, chapter 5.

Index